WELCOME TO
HOW IT WORKS
THE SCIENCE OF MEMORY

We all have memories. From our earliest recollections of childhood to what we did last night, our memories make us who we are, enabling us to chart our life to date, recall pivotal moments and plan for the future. Yet how much do we actually know about memories? How and why do they form? How many types of memory are there? Why do we have 'false' memories? Is it possible for memories to be passed from one generation to the next? In this book you'll discover the answers to all of these questions and many more. You'll delve inside the inner workings of the brain and the different regions responsible for creating and storing memories. Then it will be time to examine the effect of trauma on our ability to retrieve information before meeting a woman who can't remember the key moments in her life. Next you'll explore the science of savant minds, past lives and animal memory before stepping into the future and the mission to implant memories inside AI and the robots of tomorrow. Turn the page for what will be an unforgettable journey through the wonders of memory.

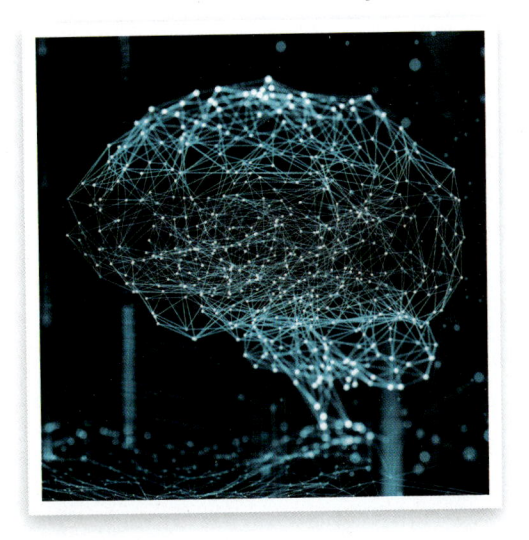

FUTURE

How It Works
THE SCIENCE OF MEMORY
EXPLORE THE MYSTERIES OF MEMORIES AND HOW THEY FORM

Future PLC Quay House, The Ambury, Bath, BA1 1UA

Bookazine Editorial
Editor **Charles Ginger**
Designer **Steve Dacombe**
Compiled by **Jacqueline Snowden & Andy Downes**
Head of Art & Design **Greg Whitaker**
Editorial Director **Jon White**
Managing Director **Grainne McKenna**

How It Works Editorial
Editor **Ben Biggs**
Designer **Duncan Crook**
Editorial Director **Tim Williamson**
Senior Art Editor **Duncan Crook**

Cover images
Shutterstock, Getty Images

Photography
Shutterstock, Getty Images, Alamy, Wikipedia
All copyrights and trademarks are recognised and respected

Advertising
Media packs are available on request
Commercial Director **Clare Dove**

International
Head of Print Licensing **Rachel Shaw**
licensing@futurenet.com
www.futurecontenthub.com

Circulation
Head of Newstrade **Tim Mathers**

Production
Head of Production **Mark Constance**
Production Project Manager **Matthew Eglinton**
Advertising Production Manager **Joanne Crosby**
Digital Editions Controller **Jason Hudson**
Production Managers **Keely Miller, Nola Cokely, Vivienne Calvert, Fran Twentyman**

Printed in the UK

Distributed by Marketforce – www.marketforce.co.uk
For enquiries, please email: mfcommunications@futurenet.com

The Science of Memory Fifth Edition (HIB5931)
© 2024 Future Publishing Limited

We are committed to only using magazine paper which is derived from responsibly managed, certified forestry and chlorine-free manufacture. The paper in this bookazine was sourced and produced from sustainable managed forests, conforming to strict environmental and socioeconomic standards.

All contents © 2024 Future Publishing Limited or published under licence. All rights reserved. No part of this magazine may be used, stored, transmitted or reproduced in any way without the prior written permission of the publisher. Future Publishing Limited (company number 2008885) is registered in England and Wales. Registered office: Quay House, The Ambury, Bath BA1 1UA. All information contained in this publication is for information only and is, as far as we are aware, correct at the time of going to press. Future cannot accept any responsibility for errors or inaccuracies in such information. You are advised to contact manufacturers and retailers directly with regard to the price of products/services referred to in this publication. Apps and websites mentioned in this publication are not under our control. We are not responsible for their contents or any other changes or updates to them. This magazine is fully independent and not affiliated in any way with the companies mentioned herein.

FUTURE Connectors. Creators. Experience Makers.

Future plc is a public company quoted on the London Stock Exchange (symbol: FUTR)
www.futureplc.com

Chief Executive Officer **Jon Steinberg**
Non-Executive Chairman **Richard Huntingford**
Chief Financial and Strategy Officer **Penny Ladkin-Brand**

Tel +44 (0)1225 442 244

Part of the
HOW IT WORKS
bookazine series

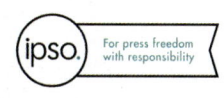

Widely Recycled

ipso. For press freedom with responsibility

CONTENTS

08 32 strange facts about memory

MEMORY BIOLOGY

- 22 Inside the human brain
- 26 How the brain remembers
- 30 Types of memory
- 34 The power of smell
- 36 Remembering dreams
- 38 Facial recognition
- 40 Childhood memories
- 42 Age-related memory loss
- 44 Active recall
- 45 Improve your memory

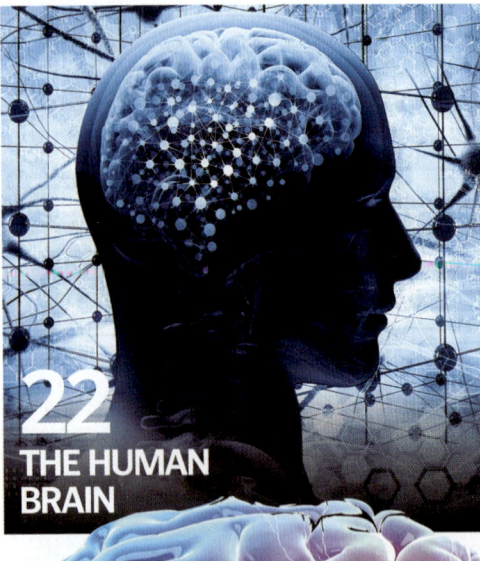

22 THE HUMAN BRAIN

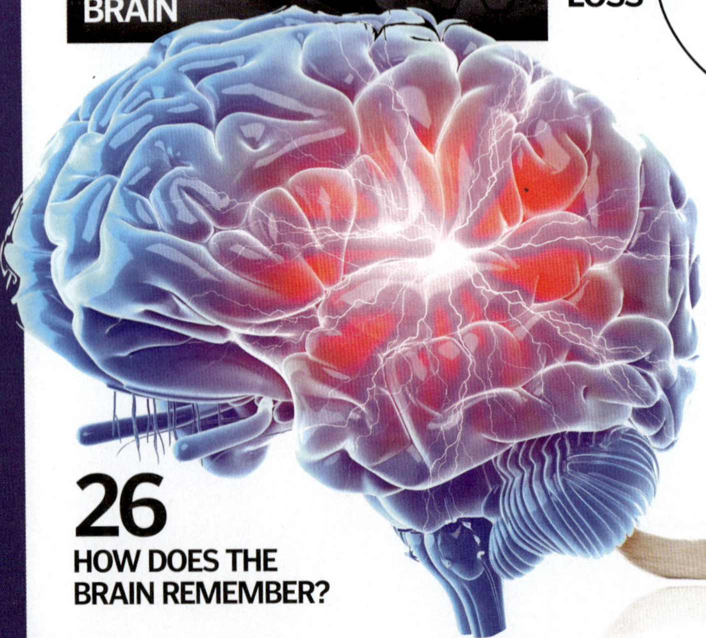

26 HOW DOES THE BRAIN REMEMBER?

54 HYPNOSIS

36 REMEMBERING DREAMS

42 MEMORY LOSS

30 TYPES OF MEMORY

TRICKS AND TRAUMA

- 48 10 mysteries of the mind
- 52 Déjà vu
- 54 Hypnosis
- 56 False memories
- 60 Eyewitness accounts
- 62 The Mandela effect
- 68 The impact of technology
- 70 Trauma and memory
- 74 Henry and Eugene
- 82 The cruellest disease

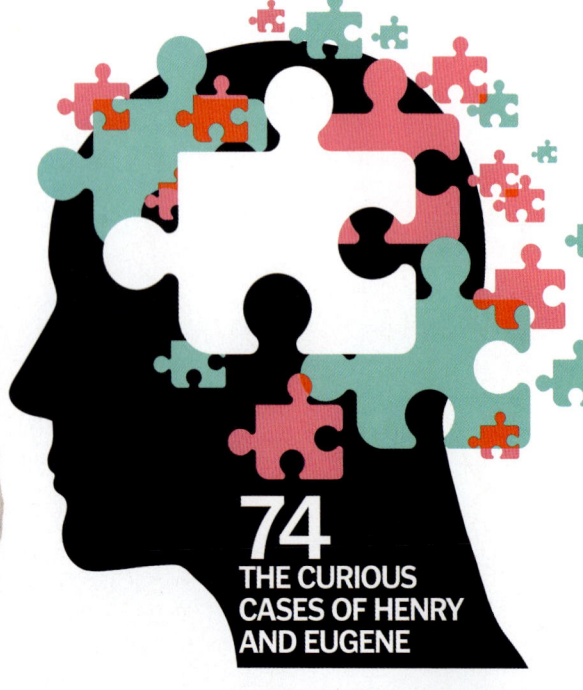

74 THE CURIOUS CASES OF HENRY AND EUGENE

006

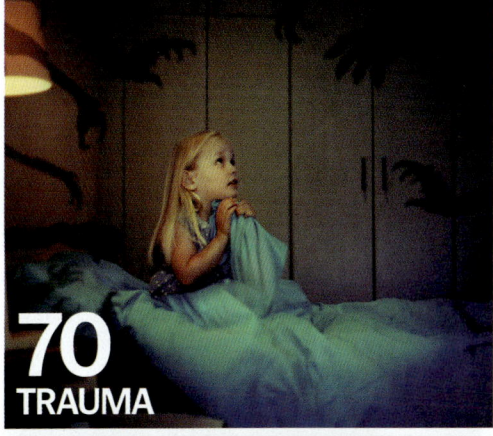

CONTENTS

40 CHILDHOOD MEMORIES

70 TRAUMA

120 DO PLANTS HAVE A MEMORY?

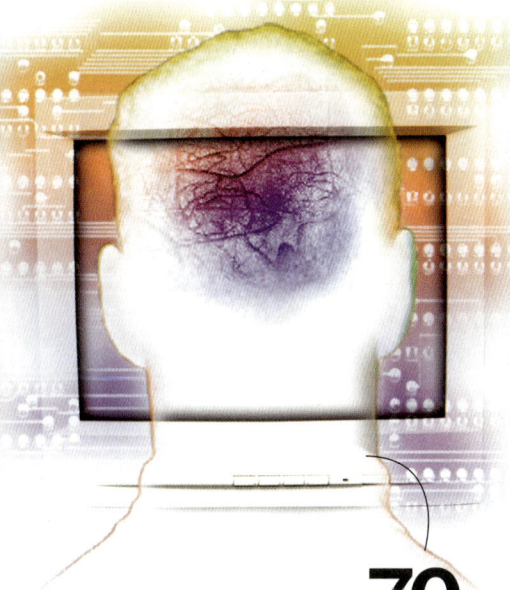

SUSIE AND THE SAVANTS

- 88 Super savants
- 96 Susie McKinnon
- 100 Photographic memory
- 102 Generational memory
- 104 Past lives

88 SUPER SAVANTS

70 HOW TECHNOLOGY CHANGES OUR MEMORY

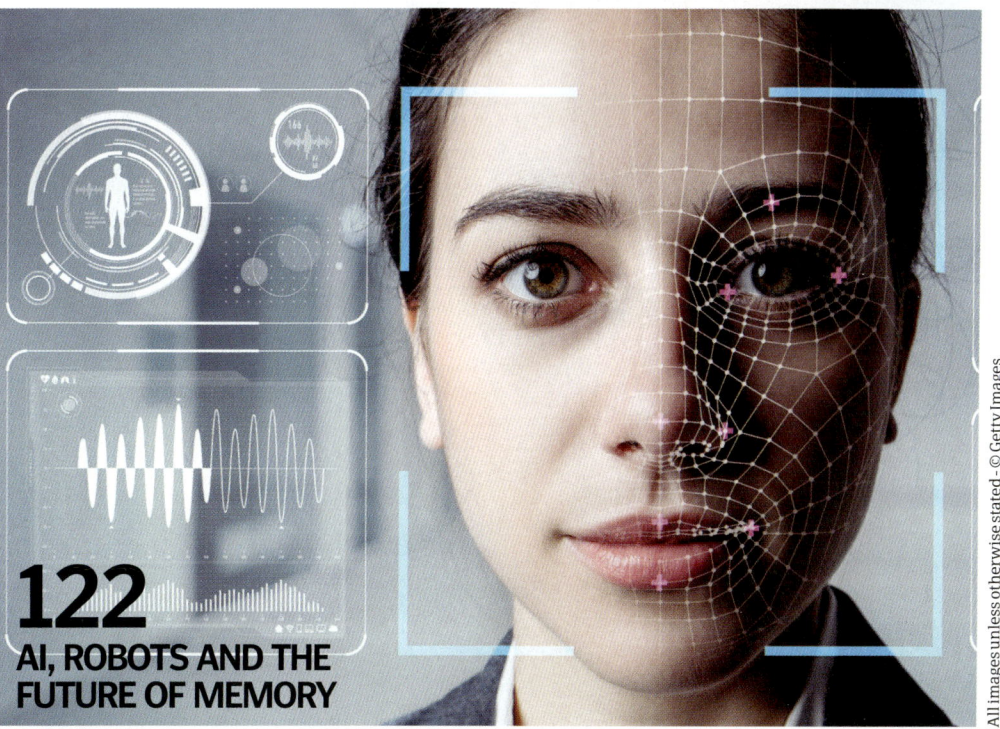

122 AI, ROBOTS AND THE FUTURE OF MEMORY

FLORA, FAUNA AND THE FUTURE

- 112 Animal memory
- 120 Plant memory
- 122 AI and the future of memory

112 ANIMALS WITH EXCELLENT MEMORY

104 PAST LIVES

007

32 STRANGE FACTS ABOUT MEMORY

MEMORY CAN BE A PLAYFUL THING – IT COLLECTS MINUTE DETAILS FROM CHILDHOOD EVENTS, YET LEAVES US WONDERING WHERE WE LEFT OUR KEYS

Words by **Bahar Gholipour and Scott Dutfield**

1 NOW WHAT WAS THEIR NAME?

Often find yourself struggling to remember names? Well don't worry – you're far from alone. While the regions in our brain that control our ability to recall faces continue to mature into our 30s, by our 20s our ability to recall names is already fading – we just don't tend to realise until we are in our 60s.

UNFORGETTABLE MEMORY FACTS

3 MEMORIES CAN LIVE ON, EVEN IF WE CAN'T ACCESS THEM

Could forgotten songs continue to live on inside our heads without us knowing? In a 2013 report of a strange case in the journal Frontiers in Neurology, researchers described a woman who had musical hallucinations of a song that she didn't recognise but others did. "To our knowledge, this is the first report of musical hallucinations of non-recognisable songs that were recognised by others in the patient's environment," the researchers wrote. The scientists said the woman had likely known the song at some point but then forgotten it. The case raises the question of what happens to forgotten memories and suggests that memories can be stored in some form in the brain that renders them accessible and yet somehow unrecognisable.

2 MIND-ERASING ACTIVITIES

Although rare, certain activities can result in temporary memory loss and brain fog known as transient global amnesia. For example, sex has been reported to cause this memory problem, with subjects forgetting the past day or so and having difficulty forming new memories afterwards. People with transient global amnesia suffer no serious side-effects, and the memory problems usually disappear in a few hours. But it's not clear how this happens, and brain scans of patients who have had this type of amnesia show no signs of damage to the brain or signs of stroke.

UNFORGETTABLE MEMORY FACTS

4 WE MAY BE PROGRAMMED TO FORGET INFANCY

Our earliest childhood memories fade, and there's likely a reason for that. Usually people don't recall any memories from their earliest years of life before the age of three or four. This is called infantile amnesia. Scientists previously thought that early memories were there but children just didn't have the language skills to verbalise them. However, recent research shows that children do make memories during their early years but forget through deliberate mechanisms. One explanation for this is that the developing brain, while generating cells, wipes out stored memories.

5 BRAIN INJURIES MAY CAUSE FORGETFULNESS

It's possible to lose memories before they even have a chance to become stored due to injuries in the brain's structures that are specifically involved in handling memory formation, maintenance and recall. Damage to these areas can result in curious forms of amnesia. In one of the most studied cases, patient H. M. lost the ability to form any new memories after his hippocampus was removed during a surgery to treat his epilepsy. Another famous case records the story of patient E. P., who had a similar fate after he suffered inflammation of the brain caused by a virus.

UNFORGETTABLE MEMORY FACTS

6 SHORT-TERM MEMORIES ARE QUICKLY FORGOTTEN

Our short-term memory is quite limited. It's thought we are only able to hold seven 'items' of memory for around 20 to 30 seconds before they are forgotten.

7 BRAIN STORAGE

It is estimated that the human brain is capable of storing anywhere between one to 1,000 terabytes of information. For comparison, a single terabyte is around 500 hours' worth of movies.

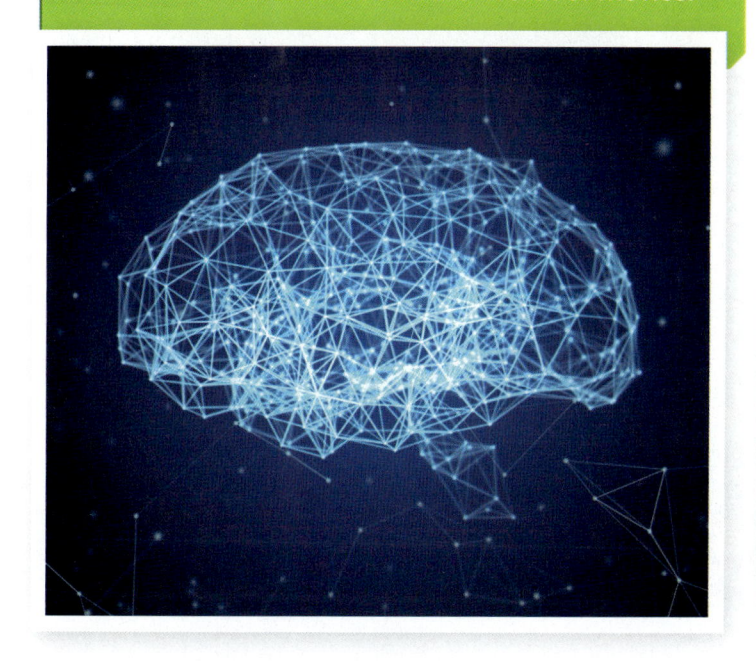

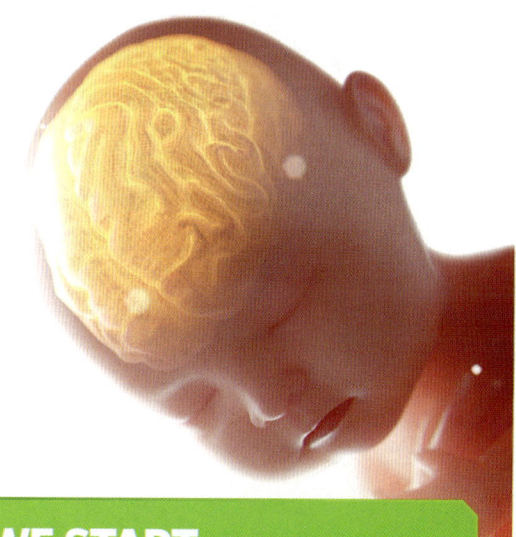

8 WE START REMEMBERING REALLY YOUNG

Research has shown that as soon as 20 weeks after conception memories begin to form during development in the womb.

011

UNFORGETTABLE MEMORY FACTS

9 TESTS ARE BEST

Research has suggested that we remember information better after being tested on it. An experiment that examined patients who were tested on a certain subject and a group that weren't found that the tested subjects were better able to recall the facts later on.

10 SMELL IS AN EXCELLENT CUE FOR MEMORIES

Often smelling a scent can send you straight back to a particular memory. This is due to the olfactory nerves' close proximity to the memory centre of the brain, the hippocampus.

11 TYPES OF LEARNERS

An estimated 65 per cent of the population are 'visual learners' and retain information from visual cues. Five per cent are what is known as an 'experimental learner' – someone who learns by doing and touching.

UNFORGETTABLE MEMORY FACTS

12 CHILDHOOD MEMORIES FADE DURING CHILDHOOD

Trying to remember what we did as children is sometimes virtually impossible. Known as 'childhood amnesia', studies have found that five to seven year olds remember around 60 per cent of early life, whereas by the age of nine that recall is down to around 40 per cent.

13 RECORD BREAKER

Prijesh Merlin achieved the world record for the most random objects memorised in 2012. He managed to recall 470 random items in the exact order in which they were read to him.

14 SAY IT OUT LOUD

A study at the University of Waterloo, Canada, found that saying information out loud increased the chance of committing it to long-term memory. The study authors dubbed their results the 'production effect' and suggested that a combination of saying the information and hearing oneself led to better recall.

UNFORGETTABLE MEMORY FACTS

15 BRAIN CONNECTIONS FORM MEMORIES

Memories are created when new connections between the brain's neurons, called synapsis, are formed. It's estimated that our brains have between approximately 100–1,000 trillion synaptic connections.

16 DRUNKEN MEMORY

It's been proven that the regular consumption of alcohol can lead to more memory mishaps in drinkers than in those who don't drink. When drinking a boozy beverage the process of transforming short-term memories into long-term ones is inhabited, and drinking can even reduce brain cell size.

17 SLEEPING HELPS STORE MEMORIES

While we sleep our brains get some much-needed time to restore and rejuvenate. This also applies to our memories as well. After 12 hours of sleep studies have shown a distinct improvement in people's ability to recall information.

UNFORGETTABLE MEMORY FACTS

18 WHAT DID I COME IN FOR?
Psychologists at the University of Notre Dame have found that people are two to three times more likely to forget what they were supposed to do after walking through a doorway.

19 TRAIN YOUR BRAIN
Studies have shown that regular exercise can improve your memory and help the areas of the brain that control memory functions to enlarge.

20 NAPPING
It might seem like a form of procrastination to nap between studying sessions, but taking a snooze has been found to improve recall. One study, featured in the journal Nature Neuroscience, found subjects could recall 85 per cent of test information after a 40-minute nap.

UNFORGETTABLE MEMORY FACTS

21 WOMEN HAVE BETTER MEMORIES
Several studies have shown that women outscore their male counterparts in memory tests, in particular when compared during middle age (between 45 to 55 years old).

22 LEFTIES HAVE BETTER RECALL
In a 2011 study comparing the memory capabilities of both left-handed and right-handed people, researchers found that lefties had a greater ability to recall their lives.

23 STARTING TO FORGET
The age at which we begin to naturally lose the sharpness of our memories was previously believed to occur at around 60. However, it's been found that this may happen earlier, at just 45.

24 CAFFEINATED MEMORIES

A study from Johns Hopkins University, USA, concluded that a dose of 200 milligrams of caffeine can enhance the consolidation of our memories. Researchers conducted a trial that involved participants who did not regularly eat or drink caffeinated products consuming a placebo or a 200-milligram caffeine tablet five minutes after studying a series of images. To ensure the test was fair, salivary samples were taken from the group before they took the tablets to measure their caffeine levels. Follow-up samples were then taken one, three and 24 hours after. On the following day the two groups were tested on their ability to recognise images from the previous day's session. Some of the visuals were the same as those previously seen, while some were new and some were similar but not the same. More members of the caffeine group were able to correctly identify the new images compared to the non-caffeine subjects.

25 BELIEVING HELPS YOU REMEMBER

A 2003 study published in The Journals of Gerontology discovered that simply believing that you have a good memory may actually help to improve your ability to recall information and events. Assuring yourself that you can enhance your memory will make you more likely to work regularly to do so.

26 EAT SOMETHING SPICY

Studies have shown that eating spicy food, particular spices used in Mexican cuisine, can help give your memory a boost. The same aid to memory has been found after eating coriander.

UNFORGETTABLE MEMORY FACTS

27 CLOSE YOUR EYES AND REMEMBER
A study published in the academic journal Legal and Criminal Psychology found that study participants who closed their eyes while trying to recall details of a film got 23 per cent more correct answers.

28 RAPID FORMATION
Within the adult hippocampus, the memory headquarters of the brain, 700 new neurons are born everyday, resulting in an annual turnover of 1.75 per cent.

29 REMEMBERING THE GOOD OLD DAYS
A phenomenon called the 'reminiscence bump' occurs over the age of 40, whereby there is a jump in the number of memories adults are able to recall from their adolescence.

UNFORGETTABLE MEMORY FACTS

30 LARGE VOCABULARY

The average person can remember between 20,000 and 35,000 words, with 10,000 of those words acquired by the age of eight.

31 IS LOVE AT FIRST SIGHT REAL?

Love at first sight might not be as real as we once hoped. A study at Northwestern University, USA, found that our memories can be corrupted by our current feelings, thus reframing and editing our memories.

32 GOOD VERSUS THE BAD

According to a study from 1930, after asking study participants to recall events throughout their lives and repeat them a week later, 42 per cent of the good memories were forgotten compared to 60 per cent of the bad ones.

MEMORY BIOLOGY

22 INSIDE THE HUMAN BRAIN

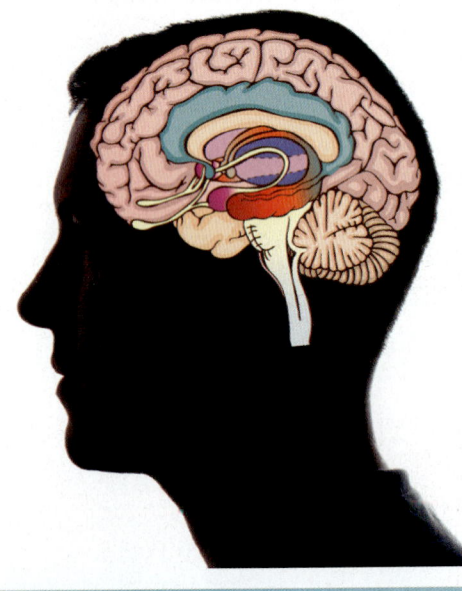

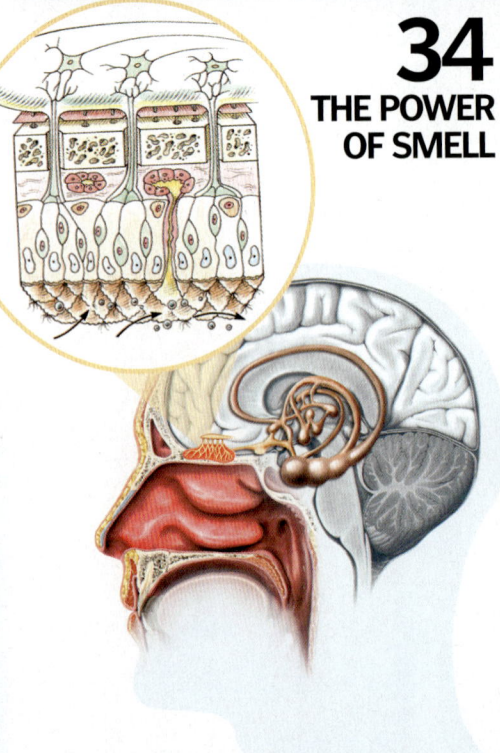

26 HOW THE BRAIN REMEMBERS

30 TYPES OF MEMORY

34 THE POWER OF SMELL

36 REMEMBERING DREAMS

MEMORY BIOLOGY

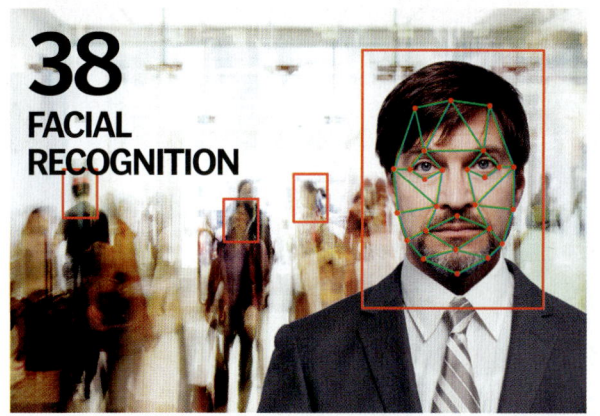

38 FACIAL RECOGNITION

"Memory doesn't merely let us memorise a shopping list – it allows us to have a meaningful life"

40 CHILDHOOD MEMORIES

42 AGE-RELATED MEMORY LOSS

44 ACTIVE RECALL

45 IMPROVE YOUR MEMORY

021

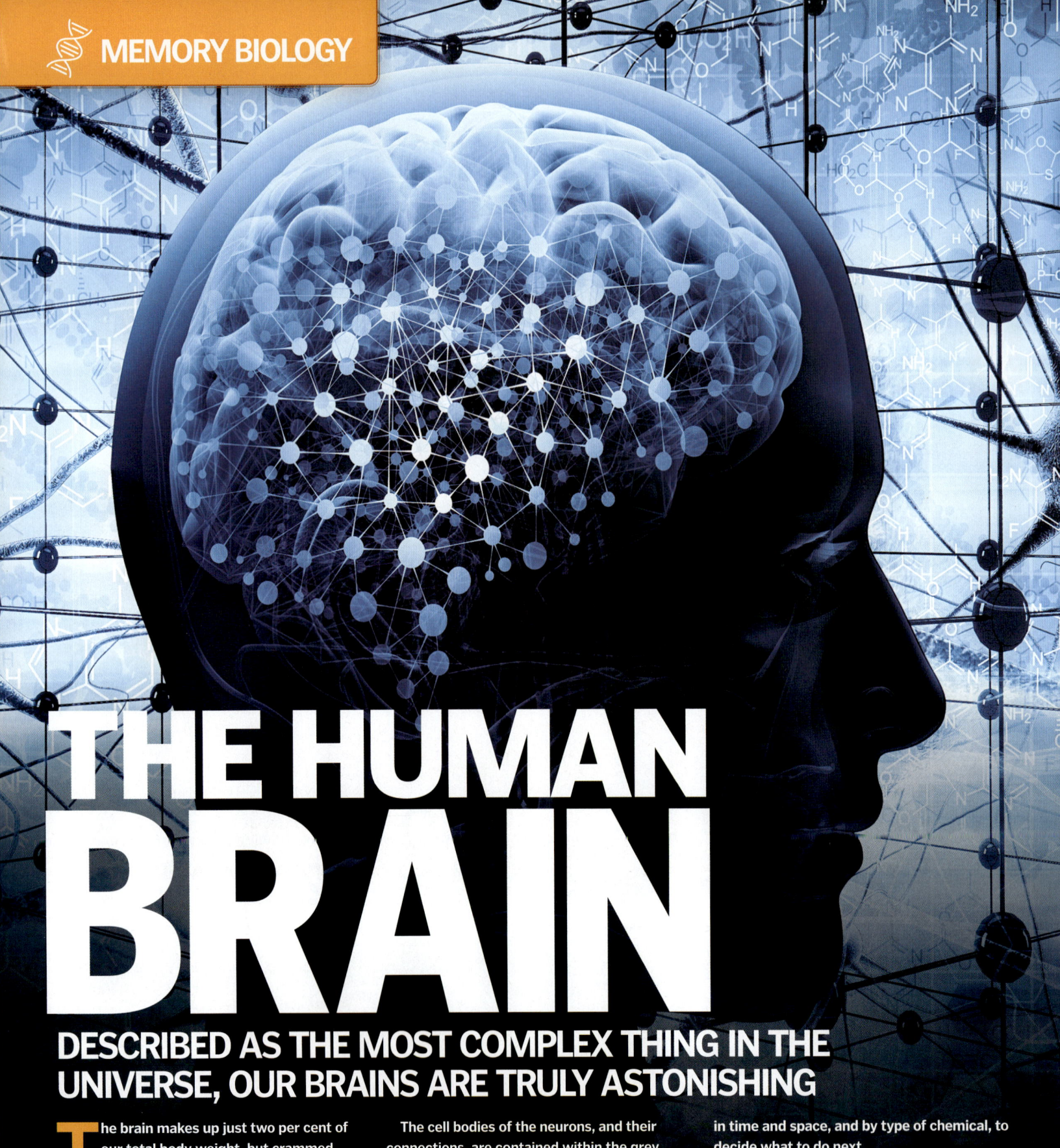

MEMORY BIOLOGY

THE HUMAN BRAIN

DESCRIBED AS THE MOST COMPLEX THING IN THE UNIVERSE, OUR BRAINS ARE TRULY ASTONISHING

The brain makes up just two per cent of our total body weight, but crammed inside are approximately 86 billion neurons, surrounded by 180,000 kilometres of insulated fibres connected at 100 trillion synapses. It's a vast biological supercomputer.

The cells in the brain communicate using electrical signals. When a message is sent, thousands of microscopic channels open, allowing positively charged ions to flood across the membrane. Afterwards, more than 1 million miniature pumps in each cell move the ions back again ready for the next impulse.

The cell bodies of the neurons, and their connections, are contained within the grey matter, which consumes 94 per cent of the oxygen delivered to the brain. Different areas are responsible for different functions, and wiring them together is a fatty network of fibres called white matter.

When a signal reaches the end of a nerve cell, tiny packets of chemical signals spill out onto the surrounding neurons. These connections, called synapses, allow messages to be passed from one cell to the next. Each neuron can receive thousands of inputs, coordinating them in time and space, and by type of chemical, to decide what to do next.

Scientists have been electrically and chemically stimulating the brain to see how it responds to different signals, recording electrical activity to map thoughts and using imaging like functional MRI to track the blood flow increases that reveal when nerve cells are firing. The cells of the brain can also be studied inside the lab. Thanks to these investigations we know more about this incredible structure than ever before, but our understanding is only just beginning. There is so much more to learn.

THE HUMAN BRAIN

BRAIN DEVELOPMENT
FROM A SINGLE CELL TO AN INCREDIBLY INTRICATE NETWORK IN JUST NINE MONTHS

Within weeks of fertilisation, neural progenitors start to form; these stem cells will go on to become all of the cells of the central nervous system. They organise into a neural tube when the embryo is barely the size of a pen tip, and then patterning begins, laying out the structural organisation of the brain and spinal cord. At its peak growth rate, the developing brain can generate 250,000 new neurons every minute. By the time a baby is born, the process still isn't complete. But, by the age of two, the brain is 80 per cent of its adult size.

Pyramidal neurons, like these, are found in the hippocampus, cortex and amygdala

20 watts Your brain is incredibly efficient, using less energy than a standard light bulb.

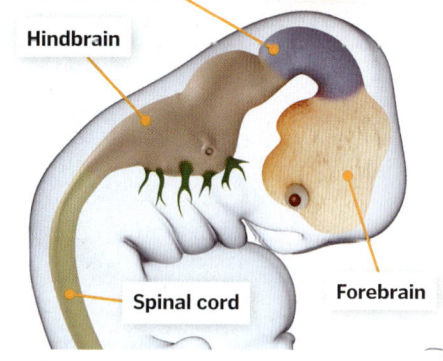

BRAIN FORMATION
THIS ASTONISHING STRUCTURE IS FORMED AND REFINED AS PREGNANCY PROGRESSES

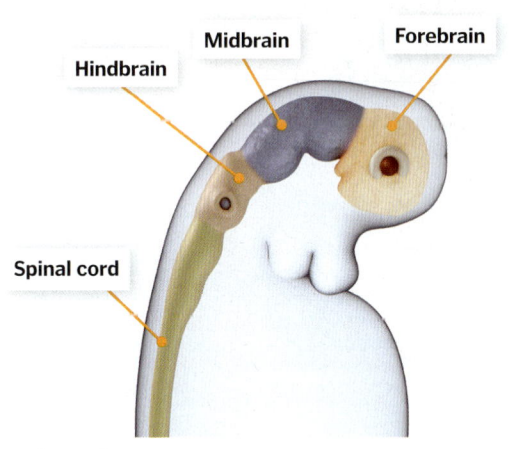

4 weeks
Brain development starts just three weeks after fertilisation. The first structure is the neural tube, which divides into regions that later become the forebrain, midbrain, hindbrain and spinal cord.

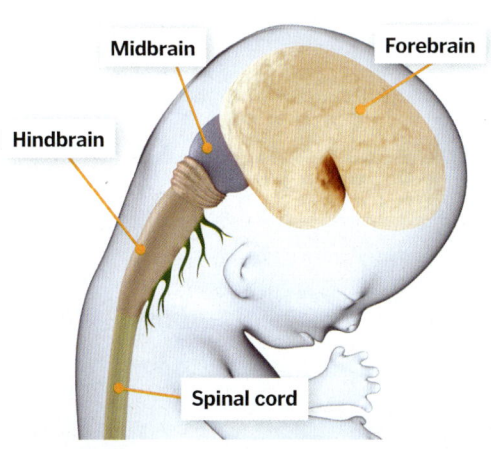

6 weeks
The pattern of the brain and spinal cord is now laid out and is gradually refined, controlled by gradients of signalling molecules that assign different areas for different functions.

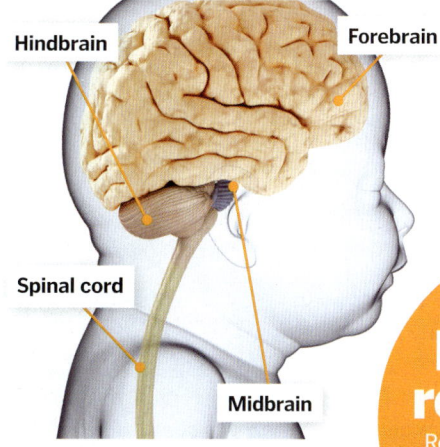

11 weeks
As the embryo becomes larger, the brain continues to increase in size and neurons migrate and organise. The surface of the brain gradually begins to fold. At this point, a foetus only measures about five centimetres in length.

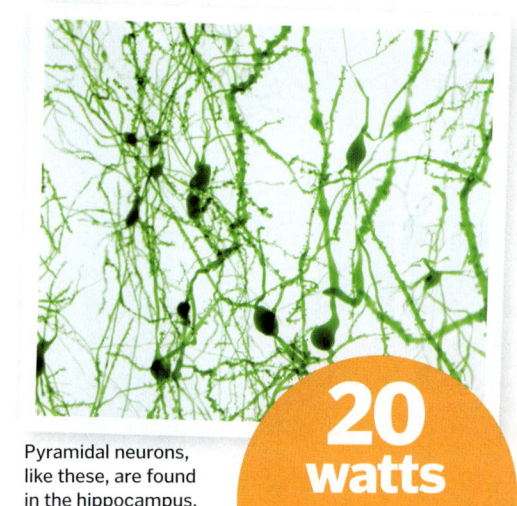

Birth
Before a baby is born, around half of the nerve cells in the brain are lost and connections are pruned, leaving only the most useful. This process continues after birth.

WHY THE BRAIN IS WRINKLED
THE BRAIN FOLDS IN ON ITSELF TO CRAM IN MORE PROCESSING POWER

The folds and pockets of our brains are a biological rarity that we only share with a few other species, including dolphins, some primates and elephants. It's a clever evolutionary adaptation that allows intelligent species to squash a huge amount of cortical tissue into a small space, allowing enormous brainpower to be crammed into our relatively small skulls.

Folding starts during the second trimester of pregnancy, creating ridges (gyri) and fissures (sulci), but the biology behind the distinctive wrinkles is stranger than you might think. The organisation of the brain is determined by complex cascades of chemical signals, but the overall shape seems to be the result of simple physics. Grey matter sits on the outside of the brain, and during development its growth rapidly outpaces the growth of white matter underneath. This puts mechanical stress on the structure, forcing the outside to buckle and curl.

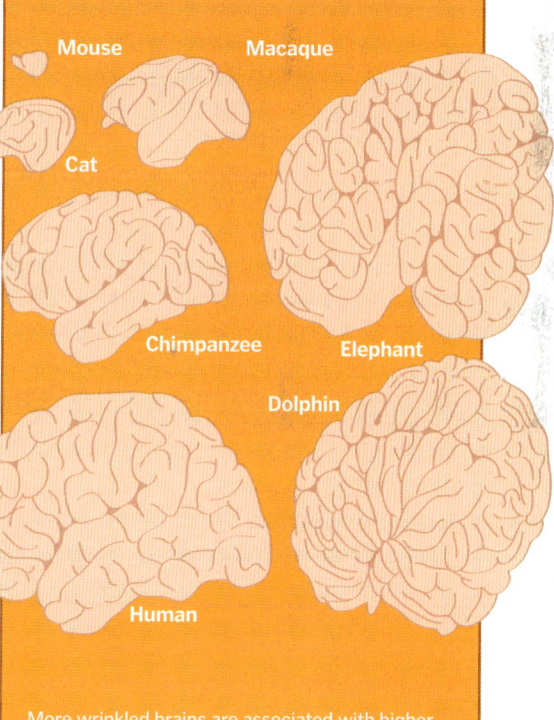

More wrinkled brains are associated with higher intelligence (brain sizes not to scale)

The brain can regenerate
Research has shown that certain areas of the adult brain can continue to produce new neurons, a process known as neurogenesis.

"Our brains contain 86 billion neurons and 180,000 kilometres of fibres"

MEMORY BIOLOGY

INSIDE YOUR BRAIN

DELVE INTO THE MAKE UP OF THE HUMAN BRAIN AND DISCOVER WHAT ITS MAJOR PARTS DO

It's a computer, a thinking machine, a fatty pink organ, and a vast collection of neurons – but how does it actually work? The human brain is amazingly complex – in fact, more complex than anything in the known universe. The brain effortlessly consumes power, stores memories, processes thoughts and reacts to danger.

In some ways, the human brain is like a car engine. The fuel – which could be the sandwich you had for lunch or a sugar doughnut for breakfast – causes neurons to fire in a logical sequence and to bond with other neurons. This combination of neurons occurs incredibly fast, but the chain reaction might help you compose a symphony or recall entire passages of a book, help you pedal a bike or write an email to a friend.

Scientists are just beginning to understand how these brain neurons work – they have not figured out how they trigger a reaction when you touch a hot stove, for example, or why you can re-generate brain cells when you work out at the gym.

The connections inside a brain are very similar to the internet – the connections are constantly exchanging information. Yet, even the internet is rather simplistic when compared to neurons.

There are around 10,000 types of neurons inside the brain, and each one makes thousands of connections. This is how the brain processes information or determines how to move an arm and grip a surface. These calculations, perceptions, memories and reactions occur almost instantaneously, and not just a few times per minute but millions. According to Jim Olds, research director with George Mason University in Virginia, if the internet were as complex as our solar system, then the brain would be as complex as our galaxy. In other words, we have a lot to learn. Science has not given up trying and has made recent discoveries about how we adapt, learn new information and can actually increase brain capability.

In the most basic sense, our brain is the centre of all inputs and outputs in the human body. Dr. Paula Tallal, a codirector of neuroscience at Rutgers University in New Jersey, USA, says the brain is constantly processing sensory information – even from infancy. "It's easiest to think of the brain in terms of inputs and outputs," says Tallal. "Inputs are sensory information, outputs

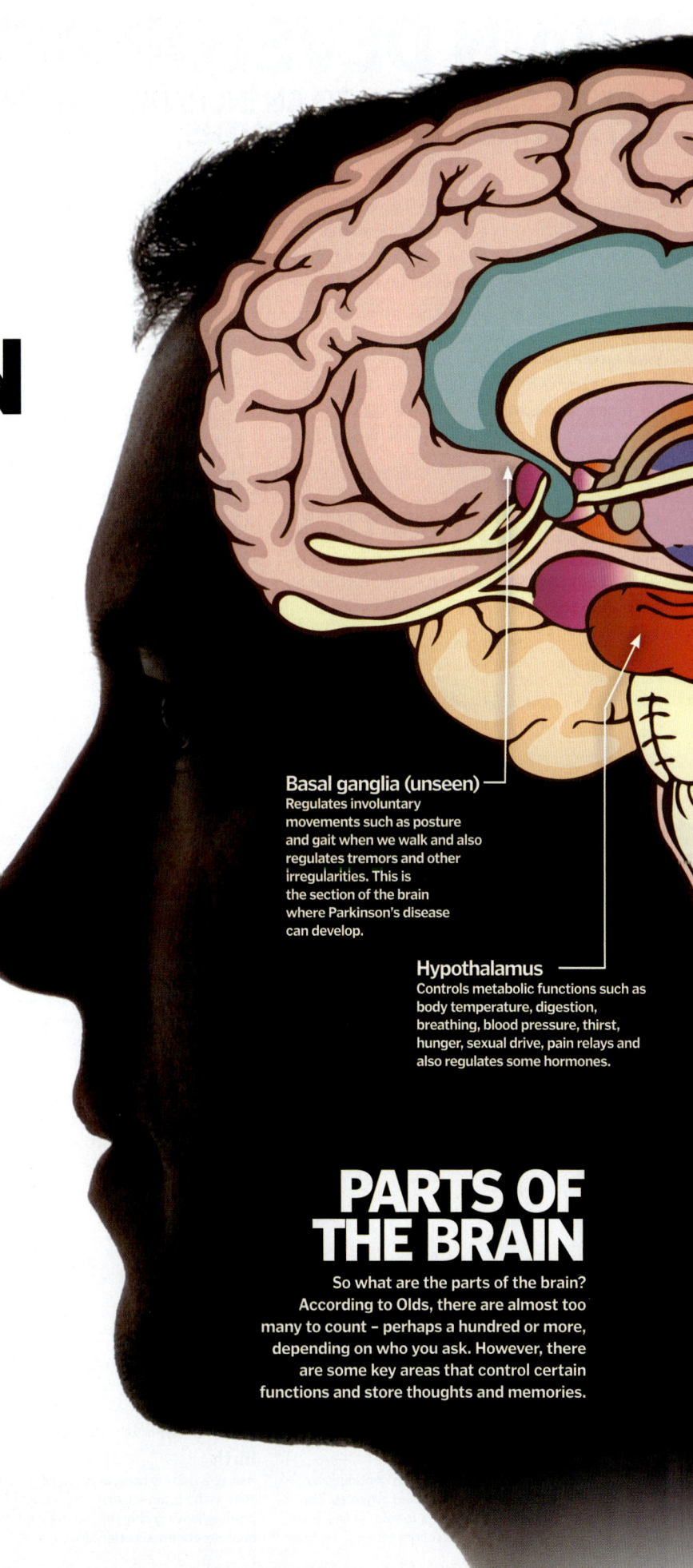

Basal ganglia (unseen)
Regulates involuntary movements such as posture and gait when we walk and also regulates tremors and other irregularities. This is the section of the brain where Parkinson's disease can develop.

Hypothalamus
Controls metabolic functions such as body temperature, digestion, breathing, blood pressure, thirst, hunger, sexual drive, pain relays and also regulates some hormones.

PARTS OF THE BRAIN

So what are the parts of the brain? According to Olds, there are almost too many to count – perhaps a hundred or more, depending on who you ask. However, there are some key areas that control certain functions and store thoughts and memories.

THE HUMAN BRAIN

FUNCTIONS OF THE CEREBRAL CORTEX

THE CEREBRAL CORTEX IS THE WRINKLING PART THAT SHOWS UP WHEN YOU SEE PICTURES OF THE BRAIN

Cerebral cortex
The 'grey matter' of the brain controls cognition, motor activity, sensation and other higher-level functions. It includes the association areas that help to process information. These association areas are what distinguishes the human brain from other brains.

Frontal lobe
This primarily controls senses such as taste, hearing and smell. Association areas might help us determine language and the tone of someone's voice.

Complex movements

Skeletal movement

Parietal lobe
This is where the brain senses touch and anything that interacts with the surface of the skin, making us aware of the feelings of our body and where we are.

Problem solving

Touch and skin sensations

Language

Receives signals from eyes

Analysis of signal from eyes

Speech
Hearing

Prefrontal cortex
This controls executive functions such as complex planning, memorising, social and verbal skills and anything that requires advanced thinking and interactions. In adults it helps to determine whether an action makes sense or is dangerous.

Analysis of sounds

Temporal lobe
This is what distinguishes the human brain – the ability to process and interpret what other parts of the brain are hearing, sensing or tasting and then determine a response.

Cerebellum
This consists of two cerebral hemispheres that control motor activity, the planning of movements, co-ordination and other body functions. This section of the brain weighs about 200g (compared to 1,300g for the main cortex).

Limbic system
This is the part that controls intuitive thinking, emotional response, sense of smell and taste.

> *"In a sense, the main function of the brain is in ordering information – interpreting the outside world and making sense of it"*

are how our brain organises that information and controls our motor systems."

Tallal says one of the primary functions of the brain is in learning to predict what comes next. In her research for Scientific Learning, she has found that young children enjoy having the same book read to them again and again because that is how the brain registers acoustic cues that form into phonemes (sounds) to become spoken words.

"We learn to put things together so that they become smooth sequences," she says. These smooth sequences are observable in the brain, interpreting the outside world and making sense of it. The brain is actually a series of interconnected 'superhighways' that move 'data' from one part of the body to another.

Tallal says another way to think about the brain is by lower and upper areas. The spinal cord moves information up to the brain stem, then up into the cerebral cortex, which controls thoughts and memories. Interestingly, the brain really does work like a powerful computer in determining not only movements but registering memories that can be quickly recalled.

According to Dr. Robert Melillo, a neurologist and the founder of the Brain Balance Centers (**www.brainbalancecenters.com**), the brain actually predetermines actions and calculates the results about a half-second before performing them (or even faster in some cases). This means that when you reach out to open a door, your brain has already predetermined how to move your elbow and clasp your hand – it may even have simulated this movement more than once, before you even perform the action.

Another interesting aspect to the brain is that there are some voluntary movements and some involuntary. Some sections of the brain might control a voluntary movement such as patting your knee to a beat. Another section controls involuntary movements, such as the gait of your walk, which is passed down from your parents.

025

MEMORY BIOLOGY

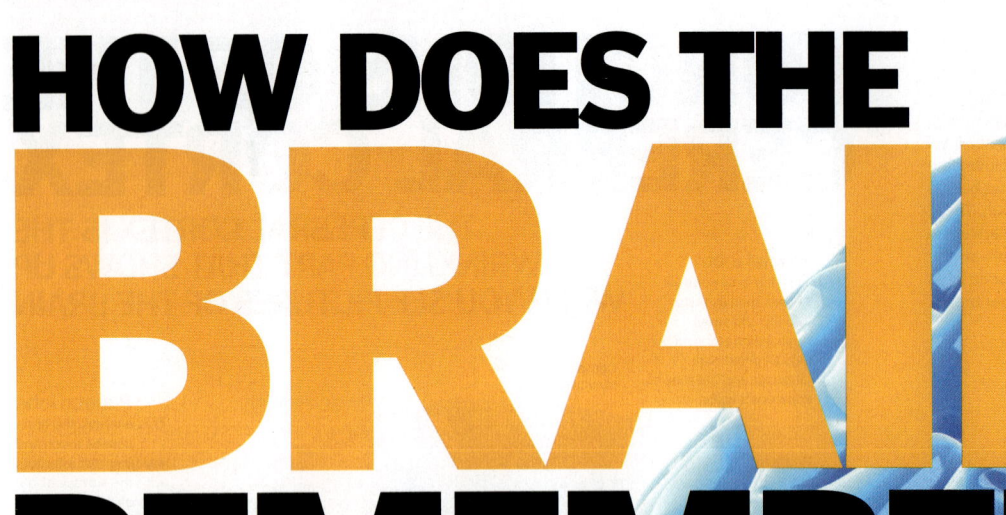

HOW DOES THE BRAIN REMEMBER?

DISCOVER HOW THIS VITAL ORGAN STORES AND UTILISES MEMORIES IN A PROCESS ESSENTIAL TO LIFE AS WE KNOW IT

Words by **Ailsa Harvey**

Where would we be without memory? Without the ability to remember significant events in our lives we would lose sense of who we are, and being unable to store information as we learn it would leave us with the permanent intelligence of a newborn.

Your memory is made up of information that's been stored in the brain and can be retrieved. It enables us to learn from experiences, build trust and understanding, develop skills through training and simply compose a thought. Memory doesn't merely let us memorise a shopping list – it allows us to have a meaningful life.

The impressive structure of the brain is so complex that scientists are constantly working to gain a better understanding of its true capabilities. Most animals have working memories, but these differ greatly – from dogs that can forget events after only two minutes to dolphins who are thought to have the best long-term memory of the entire animal kingdom.

Actions taken from recalling past memories and imagining future plans are critical to our species' survivability. The majority of other animals have adapted to store only the memories that will help them to survive. Some, such as squirrels and the chickadee bird, bury food to help them survive through harsh winters. This would be of no use at all if they weren't equipped with the memory function to relocate them. Chickadees' impressive recall allows them to find their 80,000 hidden seeds all by memory.

Our memories don't all serve a life-or-death purpose. Events that hold high significance to us are more likely to be remembered later down the line. For example, those that spark strong emotions stay with us, because they have resulted in strong connections forming inside the brain.

As soon as a memory is created it needs to be stored somewhere. Because remembering everything would quickly overload our brains, memories are first taken through a filtering process before storage. The brain takes in everything experienced by the senses. From this mass of events the ones that had the most impact on the brain are stored first as short-term memories. These memories will only be recalled for a limited time, with some of them fleeting and only lasting 20 seconds.

Those that are reused are deemed to be important, and these memories become stronger each time they are recalled. This being said, every time you retrieve the same memory of an event from your brain, it is altered slightly in some way. Because of this, no memory ever stays identical throughout life; they are more like continuously adapting reconstructions.

Witnessing something unusual is more likely to remain in your memory

> "Memory allows us to have a meaningful life"

HOW DOES THE BRAIN REMEMBER?

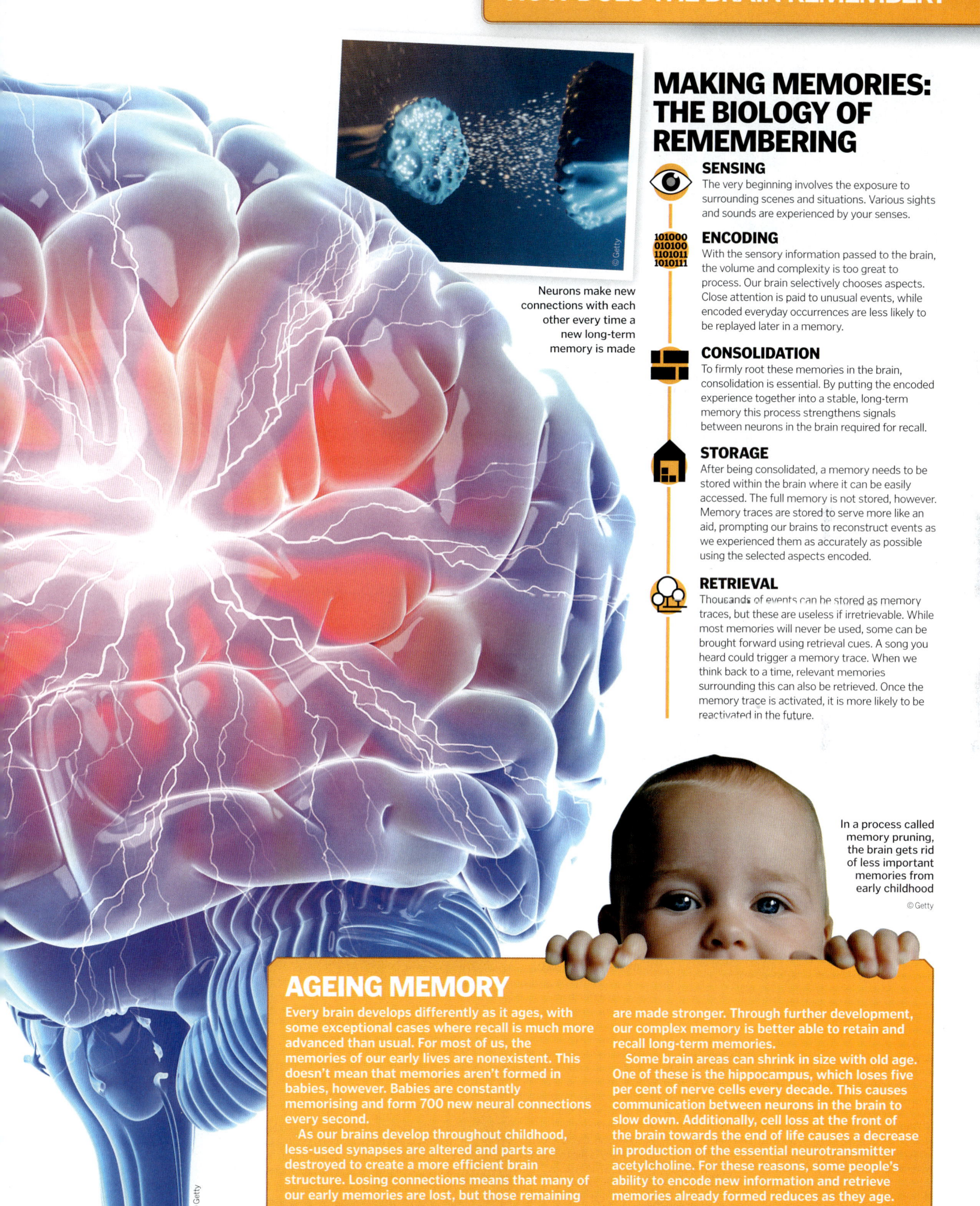

MAKING MEMORIES: THE BIOLOGY OF REMEMBERING

SENSING
The very beginning involves the exposure to surrounding scenes and situations. Various sights and sounds are experienced by your senses.

ENCODING
With the sensory information passed to the brain, the volume and complexity is too great to process. Our brain selectively chooses aspects. Close attention is paid to unusual events, while encoded everyday occurrences are less likely to be replayed later in a memory.

CONSOLIDATION
To firmly root these memories in the brain, consolidation is essential. By putting the encoded experience together into a stable, long-term memory this process strengthens signals between neurons in the brain required for recall.

STORAGE
After being consolidated, a memory needs to be stored within the brain where it can be easily accessed. The full memory is not stored, however. Memory traces are stored to serve more like an aid, prompting our brains to reconstruct events as we experienced them as accurately as possible using the selected aspects encoded.

RETRIEVAL
Thousands of events can be stored as memory traces, but these are useless if irretrievable. While most memories will never be used, some can be brought forward using retrieval cues. A song you heard could trigger a memory trace. When we think back to a time, relevant memories surrounding this can also be retrieved. Once the memory trace is activated, it is more likely to be reactivated in the future.

Neurons make new connections with each other every time a new long-term memory is made

In a process called memory pruning, the brain gets rid of less important memories from early childhood

AGEING MEMORY

Every brain develops differently as it ages, with some exceptional cases where recall is much more advanced than usual. For most of us, the memories of our early lives are nonexistent. This doesn't mean that memories aren't formed in babies, however. Babies are constantly memorising and form 700 new neural connections every second.

As our brains develop throughout childhood, less-used synapses are altered and parts are destroyed to create a more efficient brain structure. Losing connections means that many of our early memories are lost, but those remaining are made stronger. Through further development, our complex memory is better able to retain and recall long-term memories.

Some brain areas can shrink in size with old age. One of these is the hippocampus, which loses five per cent of nerve cells every decade. This causes communication between neurons in the brain to slow down. Additionally, cell loss at the front of the brain towards the end of life causes a decrease in production of the essential neurotransmitter acetylcholine. For these reasons, some people's ability to encode new information and retrieve memories already formed reduces as they age.

MEMORY BIOLOGY

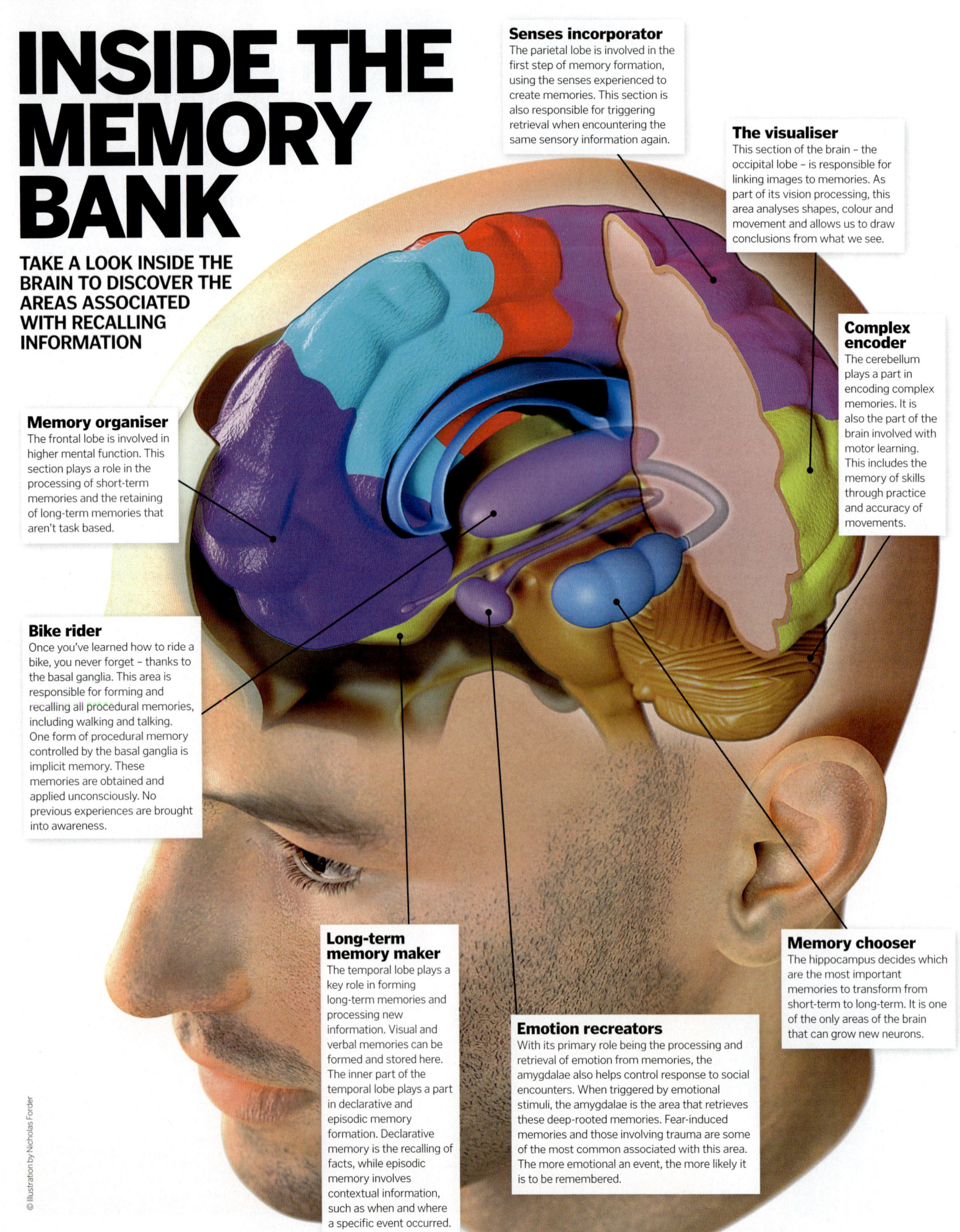

INSIDE THE MEMORY BANK

TAKE A LOOK INSIDE THE BRAIN TO DISCOVER THE AREAS ASSOCIATED WITH RECALLING INFORMATION

Senses incorporator
The parietal lobe is involved in the first step of memory formation, using the senses experienced to create memories. This section is also responsible for triggering retrieval when encountering the same sensory information again.

The visualiser
This section of the brain – the occipital lobe – is responsible for linking images to memories. As part of its vision processing, this area analyses shapes, colour and movement and allows us to draw conclusions from what we see.

Complex encoder
The cerebellum plays a part in encoding complex memories. It is also the part of the brain involved with motor learning. This includes the memory of skills through practice and accuracy of movements.

Memory organiser
The frontal lobe is involved in higher mental function. This section plays a role in the processing of short-term memories and the retaining of long-term memories that aren't task based.

Bike rider
Once you've learned how to ride a bike, you never forget – thanks to the basal ganglia. This area is responsible for forming and recalling all procedural memories, including walking and talking. One form of procedural memory controlled by the basal ganglia is implicit memory. These memories are obtained and applied unconsciously. No previous experiences are brought into awareness.

Long-term memory maker
The temporal lobe plays a key role in forming long-term memories and processing new information. Visual and verbal memories can be formed and stored here. The inner part of the temporal lobe plays a part in declarative and episodic memory formation. Declarative memory is the recalling of facts, while episodic memory involves contextual information, such as when and where a specific event occurred.

Emotion recreators
With its primary role being the processing and retrieval of emotion from memories, the amygdalae also helps control response to social encounters. When triggered by emotional stimuli, the amygdalae is the area that retrieves these deep-rooted memories. Fear-induced memories and those involving trauma are some of the most common associated with this area. The more emotional an event, the more likely it is to be remembered.

Memory chooser
The hippocampus decides which are the most important memories to transform from short-term to long-term. It is one of the only areas of the brain that can grow new neurons.

© Illustration by Nicholas Forder

HOW DOES THE BRAIN REMEMBER?

TIPS FOR A BETTER MEMORY

How you can improve your memory by making changes in your day-to-day life

PAY ATTENTION
To move information from your short- to long-term memory, paying attention and taking the time to understand information helps. Neural circuits that help build long-lasting memories work best when we focus on our surroundings. Neurotransmitters released when we are being attentive target the areas of the brain involved in processing visuals.

GET ENOUGH SLEEP
For memory consolidation to take place, your body needs sleep. While you are asleep connections in the brain can strengthen and information can pass into more permanent and efficient regions of the brain. Research shows that when information is learned before sleeping it is remembered better.

STIMULATE YOUR BRAIN
Testing your cognitive ability has been found to reduce early symptoms of memory loss. By taking part in brain games, your frontal lobe enhances its ability to split your attention between mental tasks. Keeping your brain used to memorising and keeping neuron connections activated can increase efficiency.

TRY MEDITATING
Mindfulness has been proven to enhance the abilities of your working memory. This is where new information is temporarily held. Most adults are able to hold around seven items in their working memory, but meditation is thought to strengthen it and increase its capacity.

EXERCISE REGULARLY
Physical activity has been proven to have a direct impact on brain health. By regularly exercising the risk of cognitive decline is lowered. By stimulating brain growth, studies have shown that in those who regularly exercise parts of the brain key to memory production are larger.

DRINK LESS ALCOHOL
Alcohol consumption has obvious impacts on memory ability. People who drink regularly make around 30 per cent more memory mistakes in daily life than those who don't drink. Alcohol works to prevent the transfer of short-term memories into long-term and even reduces the size of brain cells. After a heavy night of drinking it's possible to have no memory of events. This is due to a memory-affecting chemical in the brain called glutamate, which is extremely susceptible to alcohol.

MEMORY BIOLOGY

TYPES OF MEMORY

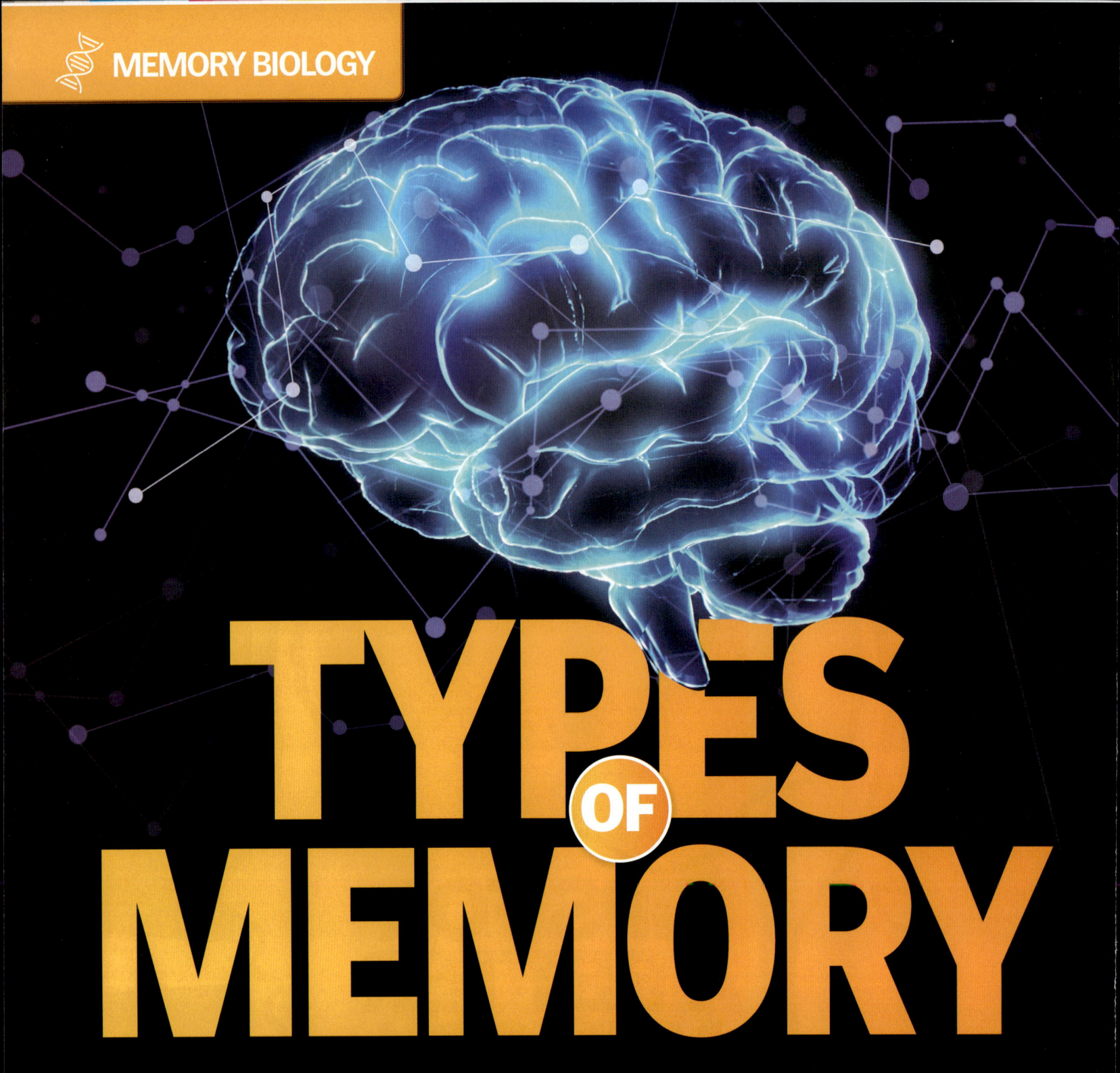

YOUR BRAIN IS SO MUCH MORE THAN A HARD DRIVE FOR INFORMATION

Words: **Laura Mears**

Your brain collects, catalogues and files data constantly, and your experiences affect everything that you do. It takes seven different types of memory to complete even the simplest of tasks.

Imagine that your phone starts to buzz as you're reading this article. Before the sound reaches your conscious attention, you have to filter it from every other sensory signal you're receiving. That takes time, so your sensory memory records a short tape of the audio to give you time to process it.

You check the sound against your long-term memory, accessing a subsystem called declarative memory to find out what it means. You then decide that it's important enough to focus on. So, you pick up your phone and unlock the screen. This uses procedural (muscle) memory – you've practised this action so much that you can do it without thinking.

You take a look at the screen and see a message from a friend. You use your short-term memory to keep track of sentences as you read and your semantic (fact) memory to work out what the words mean. They're asking you to go out tonight.

You're not sure if you're free, so you hold the time in your short-term memory while you check your diary. Then, to work out whether you want to go, you access your episodic memory, which stores information about your past. You recall old memories of your friend, make your choice, and access your procedural and semantic memories to send a reply. The simple action of replying to a friend's message takes a lot of brain power.

TYPES OF MEMORY

SENSORY MEMORY

Sensory memories are a bit like echoes, repeating the last thing you saw, heard or felt on a temporary loop. They're the reason you can work out what someone was saying even when you weren't really listening.

You only have the capacity to pay attention to around one per cent of incoming sensory information at any one time, so your brain stores the excess data in a temporary queue. It organises sensations into chunks called 'events', breaks those events into even smaller chunks called 'segments', and keeps a detailed record of every sight, sound, taste, smell and feeling happening right now.

Impressions left in the sensory memory are vivid, but they last just fractions of a second before they fade away. This gives the brain just enough time to work out which information to pay attention to and which to ignore.

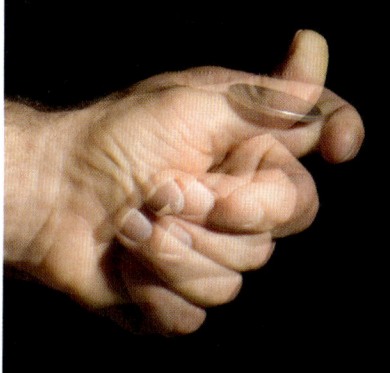

Sensory memory holds temporary echoes of incoming information

SHORT-TERM MEMORY

Short-term memory is the brain's notebook. It's a powerful problem-solving tool, but its time and space are limited. Keeping information here takes effort and attention and only happens when we're conscious. Short-term memory can only hold around seven items, give or take two, and the information fades fast.

Rehearsing short-term memories can help them to stick around for longer, and grouping information into chunks can boost memory capacity. But storage isn't really what short-term memory is for. It's much more useful as a work pad for information manipulation.

Short-term memory forms a part of working memory – the brain's rapid problem-solving system. This is the part of the memory that keeps track of sentences as you're reading, divides restaurant bills, and works out a route to the shop.

When short-term memory is active, a part of the brain called the prefrontal cortex lights up. The left side activates when you're trying to remember sounds, and the right side is triggered when you're thinking about space. These two key regions of the brain correspond to two critical workspaces used by working memory during problem-solving tasks. The first is the phonological loop – the voice inside your head – which stores audio data. The second is the visuospatial sketchpad – the mind's eye – which stores images and mental maps.

Working memory is also supported by a third short-term memory space called the episodic buffer. This area holds information from the other senses and keeps track of time. They might have small capacity, but these memory stores each have a crucial role to play in helping us to process information rapidly.

Short-term memory is the notebook for working memory

MEMORY BIOLOGY

LONG-TERM MEMORY

Short-term memories are quick and easy for the brain to make, but as soon as you lose your focus, the information disappears. Long-term memories solve this problem by storing the data permanently inside networks of connected brain cells. These networks, called ensembles, can recreate the electrical traces left by your experiences, allowing you to recall facts, events and skills on demand.

Long-term memory has two major parts: one contains information that you can access consciously (declarative memory) and the other contains instructions that you can access without thinking (procedural memory). Both types depend on a phenomenon called synaptic plasticity.

Synapses are the connections between brain cells, and your brain has around 100 trillion of them. They allow messages to pass from one cell to the next using chemical signals called neurotransmitters, but they aren't all equal; the strength of each synapse depends on how often it's used. This means that the brain is rewiring all the time, constantly adjusting the strength of existing synapses, creating new connections and pruning old ones away.

Strong connections between networks of brain cells make it easier and faster for the brain to activate specific circuits. The more you practice a skill, or repeat a fact, the stronger the connections become. But repetition isn't always needed; it's possible to store long-term memories after only one exposure. It helps if the new information links to something you already know, because many of the connections needed to store the memory are already in place.

Long-term memory is the brain's permanent hard drive

DECLARATIVE (EXPLICIT) MEMORY

Declarative memory is the closest thing to a hard drive in the human brain. It looks after the long-term memories that you can access with your conscious mind, and it is comprised of two parts. Semantic memory stores facts and episodic memory stores events.

Declarative memory is sometimes called 'explicit' memory because it contains specific knowledge: what, when, where, who and why. Long-term storage for declarative memories happens in a part of the brain called the neocortex.

Declarative memory holds both general facts and specific experiences

To get there, memories have to go via the hippocampus, which acts like a gatekeeper, deciding which information is important enough to store away. The hippocampus makes these decisions in collaboration with the brain's emotional centre, the amygdala.

The hippocampus can update and edit memories as new information comes in, but it rarely keeps a perfect record. Ask ten people what they saw at the scene of a road traffic accident and they'll all give a slightly different account.

EPISODIC MEMORY

Episodic memories can be as simple as what you had for breakfast last Monday, or as complex as a complete re-run of your first day at school. They are the brain's autobiographical storage system, holding specific information about events from the past.

The hippocampus only gets one shot at encoding episodes into long-term storage, so it relies on other brain areas for support. This includes the amygdala, which attaches emotions to memories: the stronger the emotional connection – good or bad – the harder the episodic memory is to forget. This is the reason that the most vivid episodic memories are often emotionally charged.

Other episodic memory helpers include the medial temporal lobe and the prefrontal cortex, areas that both keep memories stored in the correct order.

You can recall episodic memories at will, but they can also resurface unexpectedly during daydreaming. This plays a crucial role in future planning, allowing you to use past experiences to prepare for what might happen next.

Episodic memory stores your life experiences

TYPES OF MEMORY

SEMANTIC MEMORY

Semantic memory allows you to recognise a birthday cake

Semantic literally means 'relating to language or logic'; it's the part of the memory that handles knowledge. It looks after both specific facts, like the date of a friend's birthday, and general ideas, such as balloons.

This is the memory store that allows you to recognise a birthday cake and know that you should light the candles, even though you've probably never seen that exact cake before. Storing data into semantic memory takes time and repetition. You don't need to remember a specific party to know what a birthday cake is.

To store semantic memories, the hippocampus needs to lay down the facts, without all the extra details about where you were, or how you felt, when you learnt them. The more times you encounter the same information, the easier it is for the hippocampus to extract the facts and store them as standalone memory chunks.

"It looks after facts and general ideas"

PROCEDURAL (IMPLICIT) MEMORY

Procedural memory is the reason you never forget how to ride a bike. Sometimes known as 'muscle memory', it's the brain's storage space for 'how-to' manuals about tasks that require sequences of actions. Memories stored here include 'how to walk', 'how to play the piano', and 'how to put words together to make sentences'. You access this information unconsciously, allowing you to complete tasks without having to think about them, thereby saving precious energy.

Procedural memory involves several parts of the brain, but some of the most important are the parietal lobes, the basal ganglia and the cerebellum. The parietal lobes make links between different types of information, coupling incoming sensory data to knowledge about muscle movement. The basal ganglia helps to decide which actions to perform. The cerebellum coordinates the movements themselves, making sure everything happens in the right order and at the right time.

MEMORY BIOLOGY

THAT SMELLS FAMILIAR
WHY DO CERTAIN SCENTS TAKE YOU STRAIGHT BACK TO THE PAST?

The nose can pick up thousands of different chemical signals, allowing us to detect millions of different smells. Some of those scents trigger powerful memories, and it's all down to the wiring in our brains.

Incoming signals from the nose arrive at the olfactory bulbs before travelling on to the pyriform cortex. This part of the brain acts as a gateway, making connections to several other brain regions. There's the orbitofrontal cortex, involved in decision-making; the amygdala, the brain's emotional centre; the hypothalamus, which links the nervous and hormonal systems together; the insula, which is involved in consciousness; the entorhinal cortex, involved in memory and navigation; and the hippocampus, the master of long-term memory storage.

These connections help us to learn where smells come from and what they mean. Then, if we encounter the same smell again, we'll instantly know how to respond. For example, the brain's threat-detection centre, the amygdala, lights up when we smell something that is unpleasant.

Smells can also trigger long-forgotten memories, often in vivid and emotional detail. These reach back into early childhood, and studies in rats suggest that they form during early development. Strong odour-linked memories may help animals to survive before their other senses are fully developed – as their eyes and ears improve, the need to remember smells becomes less important. Sensing the same scents again in adulthood can bring forgotten memories flooding back.

Certain aromas can transport us straight back to childhood

MEMORY BOOST

The link between smell and memory has got scientists wondering whether we can use scents to improve our capacity to remember. Researchers at Northumbria University conducted studies to find out what happens to our brains when we sense powerful smells. In one study they asked 180 volunteers to drink chamomile tea, peppermint tea or plain hot water. Then they tested their mood and brain function. Compared to water, chamomile tea made volunteers less attentive, while peppermint tea improved their alertness.

In a separate study, 150 volunteers went into rooms that smelled of rosemary, lavender or nothing. They were then asked to complete a task at a particular time. Rosemary improved memory but lavender made it worse, although the volunteers did feel calmer.

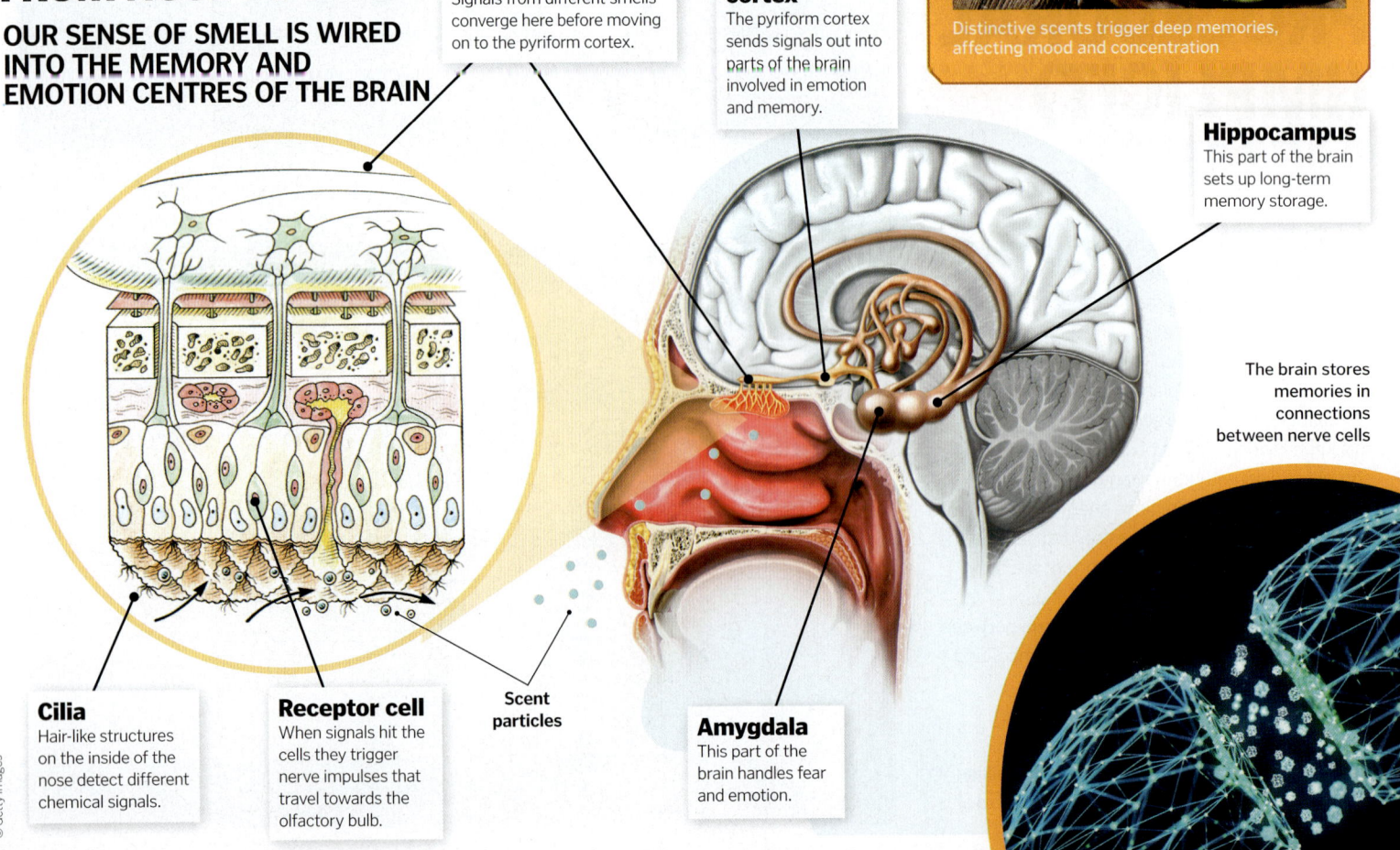

Distinctive scents trigger deep memories, affecting mood and concentration

FROM NOSE TO BRAIN
OUR SENSE OF SMELL IS WIRED INTO THE MEMORY AND EMOTION CENTRES OF THE BRAIN

Olfactory bulb
Signals from different smells converge here before moving on to the pyriform cortex.

Pyriform cortex
The pyriform cortex sends signals out into parts of the brain involved in emotion and memory.

Hippocampus
This part of the brain sets up long-term memory storage.

Cilia
Hair-like structures on the inside of the nose detect different chemical signals.

Receptor cell
When signals hit the cells they trigger nerve impulses that travel towards the olfactory bulb.

Scent particles

Amygdala
This part of the brain handles fear and emotion.

The brain stores memories in connections between nerve cells

LEARN TO BE BETTER AT EVERYTHING YOU DO

From fixing DIY disasters in a flash to mastering new skills, discover how you can get more from life! Find out the quickest and best ways to deal with household chores, set up your tech and solve your problems today.

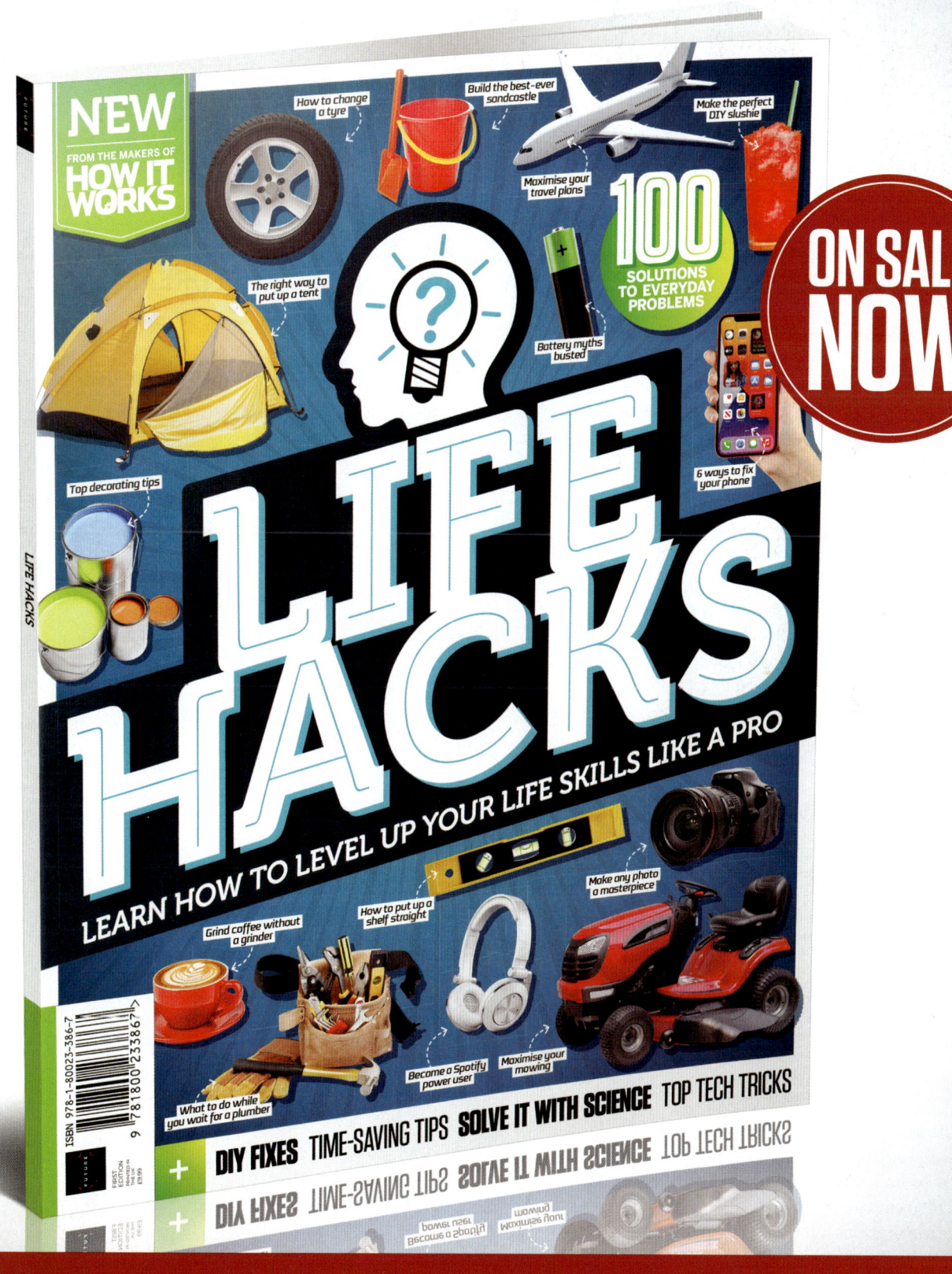

ON SALE NOW

Ordering is easy. Go online at:
magazinesdirect.com
Or get it from selected supermarkets & newsagents

MEMORY BIOLOGY

WHY CAN'T WE REMEMBER OUR DREAMS?

YOU SPEND A THIRD OF YOUR LIFE ASLEEP, A GOOD CHUNK OF WHICH INVOLVES DREAMING, YET OFTEN YOU DON'T REMEMBER ANY OF YOUR DREAMS

Words by **Bahar Gholipour**

Even on those lucky days when you wake up with a memory of the dream still floating in your mind, there's a good chance that in just a minute the memory will vanish into thin air and back to dreamland. In waking life, such a case of quickly forgetting recent experiences would surely land you in a doctor's office. With dreams, however, forgetting is normal. Why?

"We have a tendency to immediately forget dreams, and it's likely that people who rarely report dreams are just forgetting them more easily," said Thomas Andrillon, a neuroscientist at Monash University in Melbourne, Australia. It might be hard to believe that you had a dream if you don't remember anything, but studies consistently show that even people who haven't recalled a single dream in decades or even their entire lifetime, do, in fact, recall them if they are awakened at the right moment, Andrillon said.

While the exact reason is not fully known, scientists have gained some insight into memory processes during sleep, leading to several ideas that may explain our peculiar forgetfulness.

YOU ARE AWAKE, BUT IS YOUR HIPPOCAMPUS?

When we fall asleep, not all the brain's regions go offline at the same time, according to a 2011 study in the journal *Neuron*. Researchers have found one of the last regions to go to sleep is the hippocampus, a curved structure that sits inside each brain hemisphere and is critical for moving information from short-term memory into long-term memory.

If the hippocampus is the last to go to sleep, it could very well be the last to wake up, Andrillon said. "So, you could have this window where you wake up with a dream in your short-term memory, but since the hippocampus is not fully awake yet,

Our dreams can range from bizarre to mundane and blissful to terrifying

WHY CAN'T WE REMEMBER DREAMS?

It's often easier to remember dreams if they wake you up, which is why people tend to remember nightmares better than good dreams

your brain is not able to keep that memory," Andrillon explained.

While this might explain why dream memories are so fleeting, it doesn't mean that your hippocampus has been inactive throughout the night. In fact, this region is quite active during sleep and appears to be storing and caring for existing memories to consolidate them, instead of listening for incoming new experiences. Upon awakening, the brain may need at least two minutes to start its memory-encoding abilities.

In a 2017 study published in the journal *Frontiers in Human Neuroscience*, researchers in France monitored sleep patterns in 18 people who reported remembering their dreams almost every day and 18 others who rarely remembered their dreams. The team found that compared with low-dream recallers, high recallers woke up more frequently during the night. These middle-of-the-night awakenings lasted an average of two minutes for high recallers, whereas low-recallers' awakenings lasted for an average of one minute.

NEUROCHEMICAL SOUP

Our poor ability to encode new memories during sleep is also linked to changes in the levels of two neurotransmitters, acetylcholine and noradrenaline, which are especially important for retaining memories. When we fall asleep,

> "Dreams, especially mundane ones, may be deemed by the brain as too useless to recall"

acetylcholine and noradrenaline drop dramatically. Then, something strange happens as we enter the rapid eye movement (REM) stage of sleep, where the most vivid dreams occur. In this stage, acetylcholine returns to wakefulness levels, but levels of noradrenaline stay low.

Scientists have yet to work out this puzzle, but some suggest that this particular combination of neurotransmitters might be the reason we forget our dreams. The boost in acetylcholine puts the cortex in an aroused state similar to wakefulness, while low noradrenaline reduces our ability to recall our mental escapades during this time, according to a 2017 study in the journal *Behavioral and Brain Sciences*.

SOMETIMES YOUR DREAMS ARE JUST NOT MEMORABLE

Do you remember what you were thinking about this morning when brushing your teeth? Our minds wander all the time, but we discard most of those thoughts as nonessential information. Dreams, especially mundane ones, may be just like daydreaming thoughts and deemed by the brain to be too useless to recall, the late dream researcher Ernest Hartmann, who was a professor of psychiatry at Tufts University School of Medicine, wrote in *Scientific American*.

But dreams that are more vivid, emotional and coherent seem to be better remembered – perhaps because they trigger more awakening, and their organised narrative makes them easier to store, Andrillon said.

If you are intent on improving your dream recall, there are a few tricks you can try. Robert Stickgold, an associate professor of psychiatry at Harvard Medical School, suggests drinking water before bed, because it will make you wake up at night to use the bathroom. These "middle-of-the-night awakenings are frequently accompanied by dream recall," Stickgold told *The New York Times*.

When you go to bed, repeatedly reminding yourself that you want to remember your dreams may increase your chances, and so does keeping a dream journal, some studies have suggested. Upon waking up, hang on to that fragile dream memory: keep your eyes closed, stay still and replay the dream memory over and over again until your hippocampus has the chance to catch up and properly store the memory.

MEMORY BIOLOGY

NEVER FORGET A FACE

WE HAVE FACIAL RECOGNITION TO THANK FOR OUR WIDE SOCIAL CIRCLES

Facial recognition helps humans to make and maintain strong social bonds

Humans are highly social animals, relying on others for comfort, safety and happiness. We live in large groups and often have additional connections outside of our close circles, and we're able to do this in part because we possess the ability to remember and identify faces. Faces of those who are important to us – whether for positive or negative reasons – are stored in our brains ready for the next time we encounter them. Even after years apart, old friends can recognise each other across the street, while the face of a stranger walking past fades from the memory almost instantly.

A part of the brain called the fusiform gyrus is often closely associated with face and body recognition. It forms part of the ventral temporal cortex, which is a series of structures involved in the processing of high-level vision and object recognition. A small part on the lateral side, known as the fusiform face area, appears to play multiple parts in recognition, including the ability to distinguish between objects that possess very similar features.

Facial recognition isn't an ability we're born with – poor eyesight and a brain that still has a lot of developing to do mean that babies can't learn and identify faces right away. Scientists debate when they begin to recognise familiar faces, with estimates ranging from two days to two months. Self-recognition is a more complex feat, with babies usually taking around 18 months to comprehend that they are looking at their own face in a mirror. But not everyone is able to pick out a familiar face in a crowd. Some people suffer from prosopagnosia, or 'face blindness', a disorder that prevents them from remembering and recognising individual faces.

Even the ability to recognise their own face in a mirror or photo is impaired. The disorder can arise as a result of brain damage, but in some people the ability to recognise faces never develops in the first place. No cure or effective therapy exists currently, so prosopagnosics often rely on other features like voice, gait and clothing in order to tell people apart.

HAVE I SEEN YOU SOMEWHERE BEFORE?

WE'RE NOT THE ONLY ONES WHO CAN PICK OUT A FAMILIAR FACE

Many social animals can distinguish kin from strangers, and, like us, they use this ability to maintain bonds with family members and remember enemies. For creatures like paper wasps, facial recognition is a crucial tool used alongside other features like pheromones to quickly identify an intruder.

Dogs, descended from social wolves, can recognise each other and their owners, even identifying different emotions on human faces. Horses too can interpret how both their equine and human companions are feeling by looking at their expressions. Cats, on the other hand, don't seem to distinguish between human faces; instead, they rely on other cues like scent and voice tone.

Often thought of as possessing low intelligence, sheep were recently found to be able to recognise human faces. Shown a photograph of a person, they were able to pick it out when it was later displayed alongside another portrait. More impressively, they correctly selected the face of their handler without any training.

Recognising kin and even members of other species isn't uncommon in the animal kingdom, but self-recognition is a rarity. Aside from humans, only dolphins, great apes, orcas, magpies and elephants have shown that they understand who they're seeing when they look in a mirror.

Paper wasps remember faces with staggering accuracy

NEVER FORGET A FACE

"Poor eyesight and a brain that still has a lot of developing to do mean that babies can't learn and identify faces right away"

FACIAL RECOGNITION TECHNOLOGY

SOFTWARE CAN NOW PICK OUT FACES FOR US

We rely heavily on our ability to recognise faces, for both social and security purposes. It's so important that humans have developed technology that mimics our natural capabilities. Facial recognition technology employs algorithms to look for distinguishing features or analyse the size, position and shape of features on a face captured in an image or video before searching databases for known faces that match this information.

Since its development, facial recognition has been used to prevent fraudulent election votes, scan crowds for criminals, aid FBI investigations and compare travellers' faces to their passport photos. Many phones and computers offer facial recognition as a more secure alternative to a password or code. Not all facial recognition technology is applied to security-focused tasks; software can also be used to map the shape of a face so that features can be altered or animated filters applied in social media apps.

Facial recognition technology maps and measures facial features

MEMORY BIOLOGY

WHY WE LOSE OUR CHILDHOOD MEMORIES

DISCOVER WHY WE HAVE VERY FEW MEMORIES BEFORE AGE SEVEN

Words by **Baljeet Panesar**

Whether it's the day your sister was born, the day you took a toy from your brother or the first time you fell off your bike, it's likely that you were probably around the age of three or four when you had your first memory, as most of us don't tend to recall memories before this age. Some adults, however, may recall memories from the age of two, while others may not remember anything until the age of seven or eight. Even the memories that we have between the ages of three and seven are patchy; of the thousands of events we experience during our childhoods we'll only remember a few, and even less as we get older.

This phenomenon, known as 'infantile amnesia', was coined by the father of psychoanalysis Sigmund Freud, and it has remained a mystery to researchers for more than 100 years. Years ago, scientists thought that children's brains couldn't make lasting memories, but we now know that babies aged just six months can form short-term memories that last minutes and long-term memories that last weeks, while five year olds have memories that can last years. But as we reach the age of seven we tend to begin to lose our earliest memories.

During the first few years of our life, our brains make millions of connections (synapses) deep within an area known as the hippocampus. It is here that we form episodic memories – memories of events that have happened to us. Our earlier memories are more vulnerable to being forgotten, but by the age of 11 our episodic memories gradually become more adult-like and continue to mature until our late teens.

Back in 1953, a 27-year-old man named Henry Gustav Molaison (more commonly known as H.M.) underwent experimental neurosurgery to control his debilitating epilepsy. Surgeons removed Molaison's hippocampus as well as some of the surrounding area. Although the surgery reduced his seizures, Molaison could no longer form new long-term memories. He could, however, still remember facts about the world from before his operation. Before Molaison's operation it was thought that memories were stored throughout the brain. But his surgery revolutionised our understanding of memory, showing that the hippocampus is important for episodic and autobiographical memories and that different parts of the brain are responsible for different types of memory.

Having made lots of brain connections during our childhoods, we have far more in our early years than we will eventually end up with during adulthood. This process is known as 'synaptic pruning', the brain's version of natural selection. This process helps to sculpt our childhood/teenage brain into a more efficient adult version, removing weak, unused or unnecessary connections and leaving stronger and more useful connections so that you can learn more.

Even our earliest memories may not be memories but instead have been moulded from photos or stories, even if they are based on real events. But even more surprisingly, roughly 40 per cent of us have completely made up details of our first memory.

Our brains don't have the ability to store episodic memories until we're at least two years old. Perhaps your sister showed you a picture of your amazing first birthday cake or your mum told you about the time that you thought it would be funny to paint your brother. Our early memories are malleable – each time a story is told, some parts can be lost or changed and newer parts can be incorporated. As children we are much more likely than adults to form these false memories.

Intriguingly, the average age at which we have our first memories can vary by two years between different cultures. While a Maori New Zealander may recall being at a family wedding at the age of two and a half, a South Korean adult may not recall anything until the age of four and a half. The Maori's emphasis on the past and retelling family stories means that they have the earliest memories of any culture, their highly detailed narrative style influencing what their children will remember in later life.

Cultures that also value personal autonomy tend to have earlier – and more – memories than cultures that value relatedness more.

WHY WE LOSE OUR CHILDHOOD MEMORIES

WHY DO SOME CHILDHOOD MEMORIES SURVIVE?

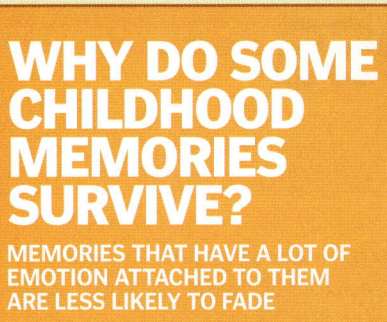

MEMORIES THAT HAVE A LOT OF EMOTION ATTACHED TO THEM ARE LESS LIKELY TO FADE

Despite the brain's pruning process, some memories can survive, although we're not completely sure why. These tend to be those that have emotional significance; perhaps you were bitten by a dog and needed stitches or you broke your arm at school. Being injured could be an evolutionary trait – those who remember the experience are (hopefully) less likely to be injured again and are therefore more likely to survive into adulthood.

In other cases, turning the incident into a story that has a sequence of events with a time and place will make it less likely to be forgotten. Perhaps you recall playing football with friends and tripping over in the school playground, landing awkwardly on your arm. Or maybe you were playing on the road after school when a neighbour's dog bit you? As we get older and develop our language skills we become more descriptive, meaning that memories become more vivid as we retell them, which makes them easier to remember.

MEMORY BIOLOGY

AGE-RELATED MEMORY LOSS
OUR MEMORY MAY NOT BE WHAT IT ONCE WAS, BUT IT'S A NORMAL PART OF AGEING

We all forget things from time to time – where we've put the keys, someone's name or why we went upstairs. If you're in your 20s, then you're probably not too concerned, but if you're in your 60s these lapses in memory can be more alarming and frustrating. We fear that they may be the start of a more serious brain disease like dementia.

As we get older, most of us will get a little more forgetful – roughly 40 per cent of over 65s experience some age-related memory loss – but this reflects normal changes in the structure and function of the ageing brain. As our brains start to shrink, some parts of the brain, like the hippocampus, a structure deep within the brain that processes our memories, and prefrontal cortex, which is responsible for language and behaviour, shrink more, and quickly than other parts. The older we get the more susceptible these parts of the brain are to ageing, taking us longer to learn, recall and recognise new information. It has been found that after the age of 40 the brain's volume decreases by five per cent each decade, and more so after the age of 70.

The brain's white matter contains nerve fibres that are surrounded and protected by a fat-rich substance called myelin. It is the nerve fibres that carry messages between brain cells, and during ageing the myelin sheath also shrinks, thereby reducing the speed that a neuron can communicate with another neuron and making it that bit more difficult to retrieve an old memory or store a new one.

The outer covering of the brain, the cerebral cortex, is also prone to thinning as a result of a reduction in synaptic connections, which help pass on messages between cells. This reduction means that it takes longer to process a memory. Research also suggests that the brain generates fewer chemical messengers, known as neurotransmitters, which include serotonin and dopamine, with advancing age, which can also lead to memory problems.

However, as we get older, the branching out of dendrites, tree-like parts of a neuron, may increase, and connections between distant brain regions can strengthen. This means that older people may have better judgment and are more rational decision-makers than their juniors. After all, with age comes wisdom.

Brain cells continue to grow even in our 90s. Brain-stimulating games such as chess may help slow down brain decline

Lifestyle, activities and daily habits have a huge impact on the brain and memory

Like humans, dogs can also experience age-related cognitive decline

DISCOVER HOW THE WONDERFUL WORLD AROUND YOU WORKS

New from the makers of How It Works magazine, 60 Second Science makes fundamental principles in physics, biology and chemistry easy to understand with clear, concise explanations, infographics and illustrations

ON SALE NOW

Ordering is easy. Go online at:
magazinesdirect.com
Or get it from selected supermarkets & newsagents

FUTURE

MEMORY BIOLOGY

THE POWER OF ACTIVE RECALL

HOW CAN WE TRAIN OUR BRAINS TO REMEMBER FACTS ON DEMAND?

Words by **Ailsa Harvey**

Every time a memory is recalled, synapses in the brain retrieve it more easily

Have you ever heard a fact that you found so impressive it never left your memory? This is because you understood its meaning and because it interested you; the more you thought about it, the more it stuck in your mind. You probably went and told your friends and family, repeating the fact. Every time you did this, you further consolidated it in your long-term memory.

This is what active recall is. It is all about actively processing the information and repeating it until it becomes a more accessible part of your memory. The issue with some facts is that we might need to learn them for an upcoming test or to impress at an interview, without it being something we are truly passionate about. Sometimes we may need to enforce these steps while studying so that we can retrieve the information at a crucial time.

Reading over the information often won't be enough. While you might understand what you are reading, unless you can demonstrate your new knowledge by recalling the information without looking at an aid, it is less likely to make the transition from your short-term memory to your long-term memory. Active recall helps to do this by strengthening memory pathways in your brain, while passive learning is more temporary.

THE KEY STEPS
FOLLOW THE PROCESS TO BUILDING UP YOUR ACTIVE RECALL

1. Take in information
Read or listen to the information you are trying to retain. Don't just skim through the words and sentences but try and make sense of them in your head.

2. Test yourself
Using only your brain and without prompts, try and remember what you have just learned. Speak aloud or write it down from memory.

3. Take your time
At the first attempt at recall, your brain's memory pathway is weak. Make a guess if necessary, but don't force memories that clearly aren't there.

4. Check yourself
When checking the information again, you may kick yourself at facts you forgot. This is because they are still in your memory. Focus on these to strengthen the neurological pathway needed to retrieve the information next time.

5. Repeat the steps
Follow the steps again multiple times. It is best to do this over the space of a couple of days, as this tests your ability to remember information even after a prolonged period of time not focusing on it.

Studies have shown that those who use active recall techniques perform 50 per cent better in test than those who learn passively

IMPROVE YOUR MEMORY

HOW TO NATURALLY IMPROVE YOUR MEMORY
FIND OUT WHICH LIFESTYLE CHOICES ARE THE BEST FOR YOUR BRAIN

I've got a memory like a sieve,' you may find yourself uttering as you fail to remember the details of an event, surrendering to forgetfulness as if it were inevitable. But what if you knew of ways to naturally improve your memory? According to science you aren't necessarily stuck with the memory you have today, and with just a few lifestyle changes you could become better at recalling information when you really need it.

As your brain is responsible for storing and retrieving all information and events, for you to bring memories forward easier you need to keep your brain at its optimum efficiency. Although cognitive ability often declines with age, through understanding what conditions your body and brain need to thrive, you can use these methods as tools to make your memory the best it can be. From general fitness to scheduled relaxation, how can making changes in your everyday life help you to remember?

A HAPPY BRAIN
KEY LIFESTYLE CHANGES TO KEEP YOUR BRAIN HEALTHY AND IMPROVE MEMORY

"For you to bring memories forward easier, you need to keep your brain at its optimum efficiency"

Exercise regularly
Physical activity has been proven to have a direct positive impact on the health of your brain. As exercise increases the oxygen in your brain, it stimulates brain growth and also reduces the risk of cognitive decline.

Brain training
Your cognitive skills get better with practice. By engaging in brain games, your frontal lobe learns to divide attention between mental tasks. This will keep neuron connections in the brain activated, improving their efficiency.

Drink less alcohol
People who drink regularly make around 30 per cent more memory mistakes than those who don't. Alcohol can prevent the ability to turn short-term memories into long-term ones as well as reducing brain cell size.

Explore meditation
Your working memory is where new information is temporarily held, and mindfulness can strengthen this. Most adults can hold seven items in their working memory, but meditation can help increase this number.

Get enough sleep
When you sleep, it enables the neuron connections in the brain to strengthen. This causes information to travel to more permanent areas of the brain to consolidate memories. So make sure that you get enough sleep!

Focus in situations
Taking the time to ensure that you understand information makes it easier to recall it later. Neural circuits that build long-term memories work more efficiently when we make the effort to actively focus on surroundings.

Keep vitamin D levels high
Links have been discovered between vitamin D levels and cognitive ability. Studies show that those who are deficient in the vitamin lose their memory quicker than those who aren't.

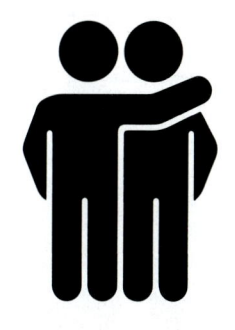

Cherish your friends
Believe it or not, simply keeping an active social life can do wonders for your memory and general brain health. Those who socialise with friends the most tend to have the lowest rate of cognitive decline.

Manage stress
Too much stress can actually damage the brain cells responsible for memory formation and retrieval. It's important to try to find the right balance between work and leisure and understand the symptoms of stress.

Maintain a healthy weight
Consuming more calories than your body needs causes weight gain as well as influencing cognitive ability. Hormones released by fat can cause parts of the brain to become inflamed.

TRICKS AND TRAUMA

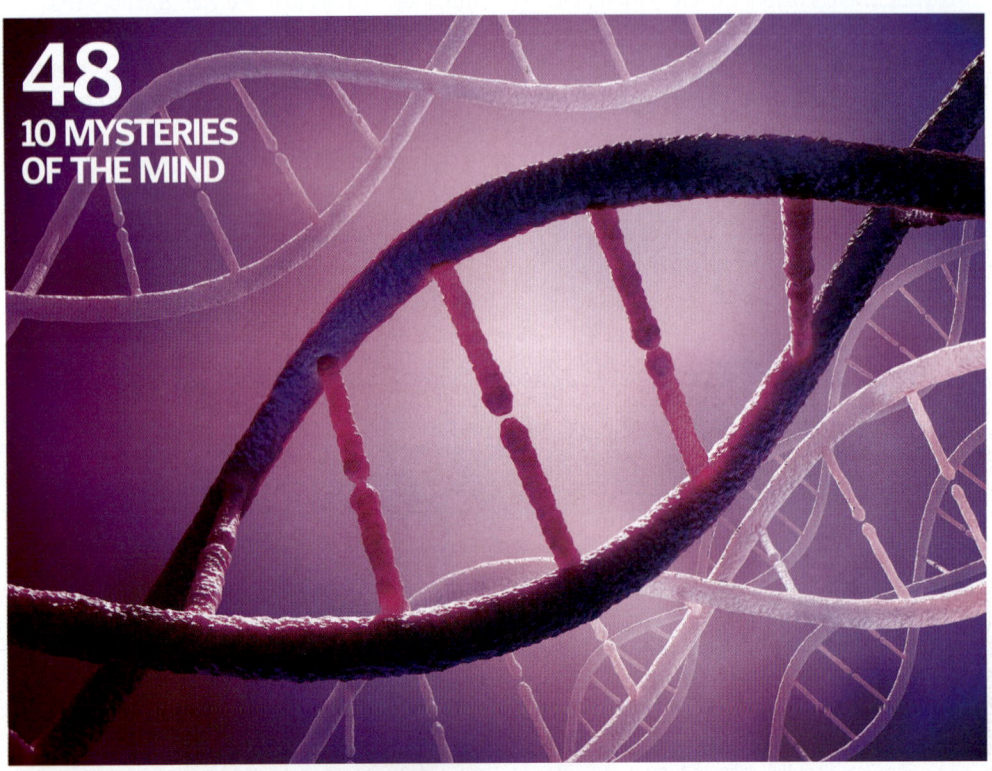

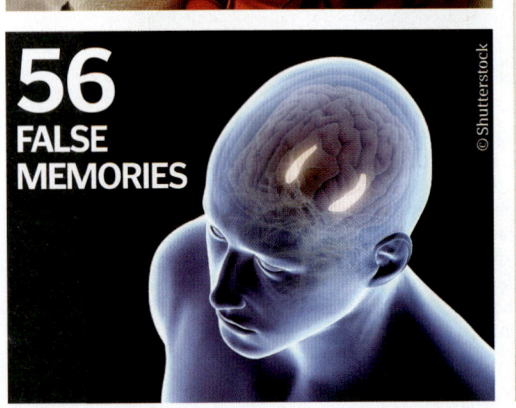

48 10 MYSTERIES OF THE MIND

52 DÉJÀ VU

54 HYPNOSIS

56 FALSE MEMORIES

60 EYEWITNESS ACCOUNTS

TRICKS AND TRAUMA

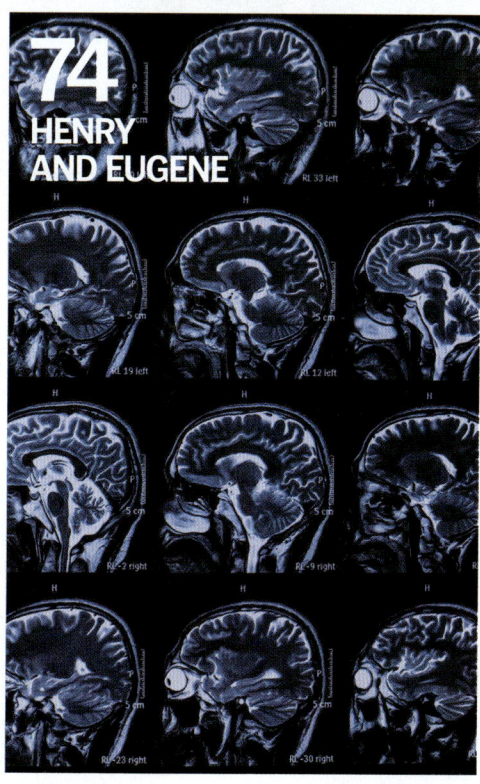

62 THE MANDELA EFFECT

68 IMPACT OF TECHNOLOGY

70 TRAUMA AND MEMORY

74 HENRY AND EUGENE

82 THE CRUELLEST DISEASE

TRICKS AND TRAUMA

10 MYSTERIES OF THE MIND

MUCH OF WHAT WE DON'T UNDERSTAND ABOUT BEING HUMAN IS SIMPLY IN OUR HEADS. THE BRAIN IS A BEFUDDLING ORGAN, AS ARE THE VERY QUESTIONS OF LIFE AND DEATH, CONSCIOUSNESS, SLEEP, AND MUCH MORE

Words: **Jeanna Bryner**

1 CONSCIOUSNESS

When you wake up in the morning, you might perceive that the Sun is just rising, hear a few birds chirping, and maybe even feel a flash of happiness as the fresh morning air hits your face. In other words, you are conscious. This complex topic has plagued the scientific community since antiquity. Only recently have neuroscientists considered consciousness a realistic research topic. The greatest brainteaser in this field has been to explain how processes in the brain give rise to subjective experiences. So far, scientists have managed to develop a great list of questions.

2 DEEP FREEZE

Living forever may not be a reality, but a pioneering field called cryonics could give some people two lives. Cryonics centres like the Alcor Life Extension Foundation in Arizona store posthumous bodies in vats filled with liquid nitrogen at bone-chilling temperatures of -320 degrees Fahrenheit (-195 degrees Celsius). The idea is that a person who dies from a presently incurable disease could be thawed and revived in the future when a cure has been found. The body of the late baseball legend Ted Williams is stored in one of Alcor's freezers. Like the other human Popsicles, Williams is positioned head-down. That way, if there were ever a leak in the tank, the brain would stay submerged in the cold liquid. Not one of the cryopreserved bodies has been revived because that technology doesn't exist. For one, if the body isn't thawed at exactly the right temperature, the person's cells could turn to ice and explode into pieces.

10 MYSTERIES OF THE MIND

3 MORTAL MYSTERY

Living forever is just for Hollywood, but why do humans age? You are born with a robust toolbox full of mechanisms to fight disease and injury, which you might think should arm you against stiff joints and other ailments. But as we age, the body's repair mechanisms get out of shape. In effect, your resilience to physical injury and stress declines. Theories for why people age can be divided into two categories: 1) Like other human characteristics, aging could just be a part of human genetics and is somehow beneficial. 2) In the less optimistic view, aging has no purpose and results from cellular damage that occurs over a person's lifetime. A handful of researchers, however, think science will ultimately delay aging for at least long enough to double people's lifespans.

4 NATURE VS. NURTURE

In the long-running battle of whether our thoughts and personalities are controlled by our genes or our environment, scientists are building a convincing body of evidence that it could be either or both. The ability to study individual genes points to many human traits that we have little control over, yet in many realms peer pressure or upbringing has been shown to heavily influence who we are and what we do.

5 BRAIN TEASER

Laughter is one of the least understood of human behaviours. Scientists have found that during a good laugh three parts of the brain light up: a thinking part that helps you get the joke, a movement area that tells your muscles to move, and an emotional region that elicits the 'giddy' feeling. But it's not known why one person laughs at your dad's foolish jokes while another chuckles while watching a horror movie. John Morreall, who is a pioneer of humour research at the College of William and Mary, has found that laughter is a playful response to incongruities – stories that disobey conventional expectations. Others point to laughter as a way of signalling to another person that this action is meant 'in fun'. One thing is clear: laughter makes us feel better.

TRICKS AND TRAUMA

6 MEMORY LANE

Some experiences are hard to forget, like perhaps your first kiss. But how does a person hold onto these personal movies? Using brain-imaging techniques, scientists are unravelling the mechanism responsible for creating and storing memories. They are finding that the hippocampus, located within the brain's grey matter, could act as a memory box. But this storage area isn't so discriminatory. It turns out that both true and false memories activate similar brain regions. In order to pull out the real memory, some researchers ask a subject to recall the memory in context, something that's much more difficult to do when the event didn't actually occur.

7 MISSION CONTROL

Residing in the hypothalamus of the brain, the suprachiasmatic nucleus, or biological clock, programs the body to follow a 24-hour rhythm. The most evident effect of circadian rhythm is the sleep-wake cycle, but the biological clock also impacts digestion, body temperature, blood pressure and hormone production. Researchers have found that light intensity can adjust the clock forward or backward by regulating the hormone melatonin. The latest debate is whether or not melatonin supplements could help prevent jet lag – the drowsy, achy feeling you get when 'jetting' across time zones.

8 PHANTOM FEELINGS

Around 80 per cent of amputees may experience sensations, including warmth, itching, pressure and pain, coming from their missing limb. People who experience this 'phantom limb' phenomenon feel sensations as if the missing limb were still there. One explanation is that the nerves in the area where the limb was severed create new connections to the spinal cord and continue to send signals to the brain. Another possibility is that the brain is 'hard-wired' to operate as if the body were fully intact, meaning that the brain holds a blueprint of the body with all of its original parts attached.

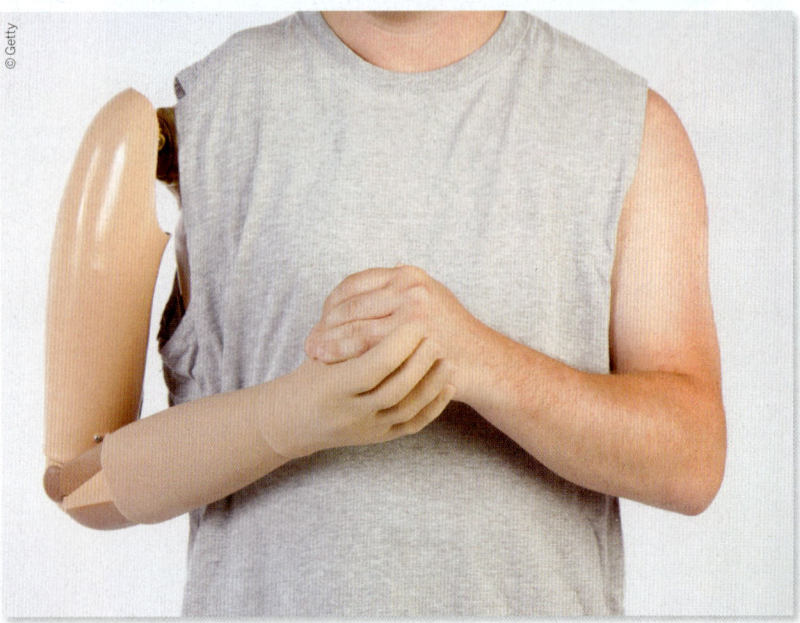

10 MYSTERIES OF THE MIND

9 SLUMBER SLEUTH

Fruit flies do it. Tigers do it. And humans can't seem to get enough of it. No, not that. We're talking about shut-eye, so crucial we spend more than a quarter of our lives at it. Yet the underlying reasons for sleep remain as puzzling as a rambling dream. One thing scientists do know is that sleep is crucial for survival in mammals. Extended sleeplessness can lead to mood swings, hallucination, and in extreme cases, death. There are two states of sleep: non-rapid eye movement (NREM), during which the brain exhibits low metabolic activity, and rapid eye movement (REM), during which the brain is very active. Some scientists think NREM sleep gives your body a break, and in turn conserves energy, similar to hibernation. REM sleep could help to organise memories. However, this idea isn't proven, and dreams during REM sleep don't always correlate with memories.

10 SWEET DREAMS

If you were to ask ten people what dreams are made of, you'd probably get ten different answers. That's because scientists are still unravelling this mystery. One possibility is that dreaming exercises the brain by stimulating the trafficking of synapses between brain cells. Another theory is that people dream about tasks and emotions that they didn't take care of during the day, and that the process can help solidify thoughts and memories. In general, scientists agree that dreaming happens during your deepest sleep, which is called rapid eye movement (REM).

TRICKS AND TRAUMA

HAVE YOU READ THIS BEFORE?
DISCOVER THE SCIENCE OF DÉJÀ VU AND THE TECHNIQUE USED TO TRIGGER IT

Around 70 per cent of us experience it, in particular those of us aged 15-25, and it can be one of the most jarring feelings: déjà vu. French for 'already seen', it has previously been linked to the theory of false memories; the idea that we can view something once and when exposed to a scene or situation that is similar our brain will respond by creating a memory that didn't really happen. However, an experiment led by psychology researcher Akira O'Connor in 2016 revealed that this might not be the case. Rather than false memory, the brain is memory checking and sending an error message, signalling what we have actually experienced versus what we think we have experienced. Around 70 per cent of us experience... wait a minute...

Déjà vu is more common in younger people, becoming less common as we age

O'CONNOR'S EXPERIMENT
HOW DID SCIENTISTS ARTIFICIALLY TRIGGER DÉJÀ VU IN THE STUDY'S VOLUNTEERS?

STEP 1
Participants were given a list of words to remember including 'bed', 'pillow', 'dream' and 'doze'; all words that are connected, in this case, to the word 'sleep'.

STEP 2
They were then asked if any of the words in the list began with the letter 'S'. Each person correctly said no.

STEP 3
Later on, the volunteers were asked if the word 'sleep' was included in the previous list of words. This prompted a feeling of déjà vu.

STEP 4
Those experiencing the chilling phenomena were scanned using functional magnetic resonance imaging (fMRI) to identify the active parts of their brain.

STEP 5
Scans revealed that the memory centre of the brain, the hippocampus, was unexpectedly not active, but the frontal areas that handle decision-making were active instead.

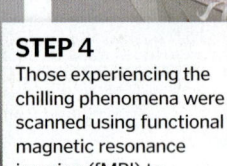

TRAPPED IN AN ENDLESS TIME LOOP
The man who has déjà vu about déjà vu

The term 'déjà vu' was coined in 1876 by French philosopher Émile Boirac

Déjà vu can be a disconcerting experience, often one that causes confusion or frustration. However, most of us are lucky enough to only get déjà vu occasionally, unlike one unfortunate man in his mid-20s who is constantly stuck inside a never-ending cycle of it. Believed to have been initiated by anxiety, this British man's condition is so severe that he avoids watching TV, listening to the radio or reading newspapers, the sense that he has "encountered it all before" proving to be completely inescapable. Scientists who studied this case of 'chronic déjà vu' found that the subject had a history of depression and anxiety, and he had once taken LSD, but otherwise he was healthy. His brain scans returned normal results, implying the cause was psychological.

The roots of déjà vu still evade science, but researchers such as Akira O'Connor believe a 'brain twitch' may be responsible.

052

EXPLORE THE HISTORY & SCIENCE BEHIND DEADLY OUTBREAKS

Learn more about some of history's most dangerous pandemics, from the Black Death to Covid-19. You'll also discover how viruses work, and see how vaccines and other breakthroughs can fight them.

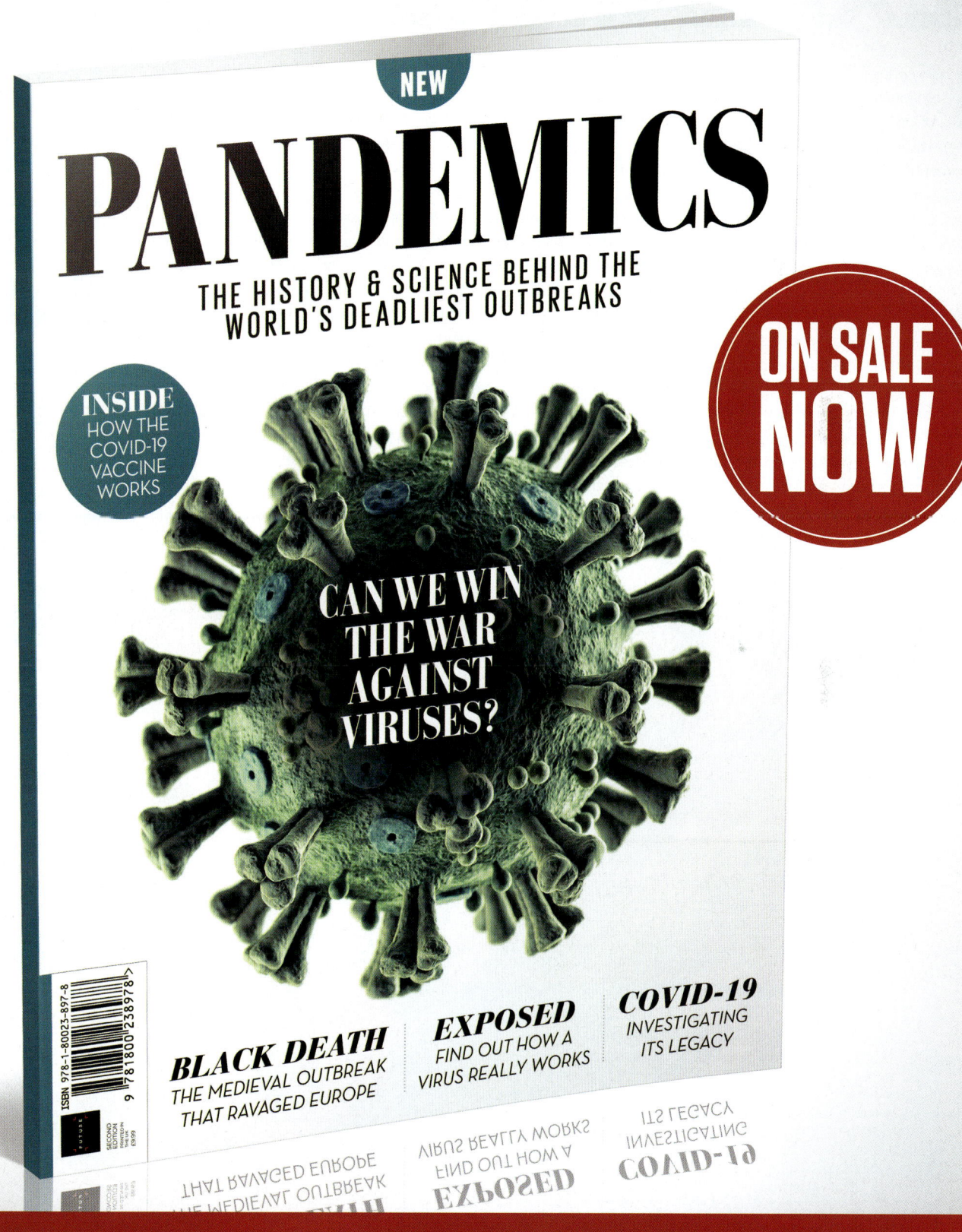

NEW

PANDEMICS
THE HISTORY & SCIENCE BEHIND THE WORLD'S DEADLIEST OUTBREAKS

INSIDE HOW THE COVID-19 VACCINE WORKS

CAN WE WIN THE WAR AGAINST VIRUSES?

ON SALE NOW

BLACK DEATH THE MEDIEVAL OUTBREAK THAT RAVAGED EUROPE

EXPOSED FIND OUT HOW A VIRUS REALLY WORKS

COVID-19 INVESTIGATING ITS LEGACY

ISBN 978-1-80023-897-8

future

Ordering is easy. Go online at:
magazinesdirect.com
Or get it from selected supermarkets & newsagents

TRICKS AND TRAUMA

HYPNOSIS EXPLAINED

SUPERNATURAL MIND CONTROL, PLACEBO EFFECT OR SOMETHING IN BETWEEN? HYPNOSIS TAKES US ON A JOURNEY INTO THE MIND…

In its simplest terms, hypnosis is a process by which someone becomes less aware of conscious thought and inhibition and more open to suggestion. Changes in the brain's neural activity can alter the subject's perceptions and emotions, enabling them to focus their thoughts and filter out distractions. One key area involved in such altered states includes the frontal lobe, which accounts for a large portion of the brain's mass and is responsible for a person's personality, emotions and long-term memory. Changing the brain's frontal lobe function in turn alters a person's subjective experience of reality as cognitive processes shift and elective actions occur without conscious volition.

Other areas of the brain that are involved with altered states include the parietal lobe, which can distort the subject's perception of space and time; the thalamus, which can induce the feeling in a subject that they're 'in a world of their own'; and the reticular formation, which receives sensory information from the outside world and determines what is important and what's not, so as to prevent us from suffering sensory overload.

Typically a hypnotist will 'induce' the subject into a highly suggestible state via techniques such as progressive relaxation or surprise. However, a formal induction isn't a prerequisite for hypnotism to succeed.

Clinical hypnosis is conducted to address both psychological and physical problems. For example, it has been used to reduce the experience of pain in severe burn victims and of women in labour and to offer relief from nausea to patients undergoing chemotherapy. Hypnosis is also used to treat those with anxiety and various phobias as well as to modify behaviour in the treatment of eating disorders and smoking cessation.

Hypnotism is a form of dissociation that works by allowing the patient to respond to suggestion while ignoring competing or incompatible stimuli. This is achieved by means of existing mental faculties. People who are hypnotised have the same physical and mental abilities that they possess in a normal state. They cannot be empowered to perform acts of superhuman strength, nor can they be forced to recall events that they never retained, such as memories of their infancy.

The human brain is capable of entering an altered consciousness whereby the subject undergoes a range of conscious experiences

WHO CAN BE HYPNOTISED?

The ability to be hypnotised falls along a normal distribution, or bell-shaped curve, with the majority of people being moderately responsive to hypnotic suggestion, and smaller numbers at the extremes, either very difficult or very easy to hypnotise. Hypnotisability – much the same as IQ – is a relatively stable quality that will remain consistent throughout adulthood.

Scientists are always searching for characteristics that will predict successful hypnotism. They have ruled out any association between hypnotisability and being 'weak willed' or gullible. Nor are people with dissociative qualities or excellent imaginations especially open to this practice. However, it does appear that people who have the ability to become completely engrossed in daydreams or music are more likely to respond to hypnosis than those who cannot.

THE STATE DEBATE

One controversy surrounding hypnosis is the state debate. While professionals on both sides of the argument agree hypnosis exists, they disagree about the way it takes place. Subscribers to the state theory believe hypnotic induction puts participants into an altered state of consciousness, which is totally discrete from normal waking reality. State theorists argue that it is this shift of state that allows for the atypical behaviour that's often observed during hypnosis.

Non-state theorists, on the other hand, maintain that the process of hypnosis moves participants along a continuum into a zone of heightened suggestibility, perhaps due to their expectations or compliance, but that the mechanisms employed during hypnosis are the same as those governing normal consciousness.

TYPES OF HYPNOSIS

WE ALL KNOW WHAT HYPNOSIS IS, BUT DID YOU KNOW THAT THERE IS MORE THAN ONE TYPE?

ERICKSONIAN

Named after American psychologist Milton H. Erickson, this type of hypnotherapy deploys indirect suggestions and storytelling to change behaviour. It is a method that is viewed as pioneering in the world of hypnosis and one that is widely adopted.

SUGGESTION

Often used to alter habits, reduce stress and treat anxiety, this method involves a hypnotherapist making suggestions while their patient is in a hypnotic state, when the subconscious is thought to be more open to guidance and new ideas.

COGNITIVE

Blending a number of hypnotherapies in order to provide a subject with a customised plan aligned with their needs and goals, cognitive hypnotherapy is used to 'update' the subconscious and bring it into line with the subject's conscious state. This type of therapy is particularly effective for curing phobias.

HYPNO-PSYCHOTHERAPY

Commonly used with subjects who have deeper problems to tackle, hypno-psychotherapy is practised by a professional trained in both hypnosis and psychotherapy. An integrative method, it is used in conjunction with another type of psychotherapy, such as mindfulness or humanist.

HYPNOANALYSIS

This approach seeks to uncover the 'trigger event' causing the subject's current issues. Unearthing the root of the problem enables the hypnotherapist to lessen the negative feelings associated with the event. It can be used alongside other types of therapeutics.

PAST LIFE REGRESSION

Adherents to this approach believe that everyone has past lives and that by exploring them issues in the subject's present life can be dealt with. It is thought to be helpful for people who feel they are stuck in life or find that certain problems continue to reoccur.

TIME LINE THERAPY

This type of therapy is predicated on the belief that our memories are stored in a sequential order (a timeline). Therapists who use this method rely on a variety of techniques to enable subjects to release detrimental beliefs and feelings that are currently holding them back, helping to treat depression and anxiety.

SOLUTION

Unlike the other types of hypnosis, solution-focused hypnotherapy sees the subject leading the sessions, with the therapist acting as a guide. This method works to unleash the inner reserves of strength within people in order to enable them to change their here and now and shape their future.

TRICKS AND TRAUMA

FALSE MEMORIES

DOES YOUR BRAIN RECORD MEMORIES LIKE A VIDEOTAPE, OR IS OUR RECOLLECTION MORE OF AN ARTIST IMPRESSION OF PAST EVENTS?

Words by **Scott Dutfield**

We are particularly precious about our memories. As records of life events in a scrapbook locked away in our minds, our memories are the biological biographies of our history and a reference guide on how to function in daily life. From remembering how to cook your family's pasta recipe, your tenth birthday party or monumental moments such as your wedding day, memories are often thought of as like a video recording of time. But what if your recall of life events isn't as accurate as you originally thought? Could 'false memories' be unwittingly infiltrating your brain, creating details of an event that never happened?

By its very definition, a false memory is one that is a fabricated or distorted recollection of an event, often surrounding fact. For example, imagine walking down the street and catching a glimpse of

American psychologist Professor Elizabeth Loftus has dedicated her career to understanding how false memories form

FALSE MEMORIES

"Could 'false memories' be creating details of an event that never actually happened?"

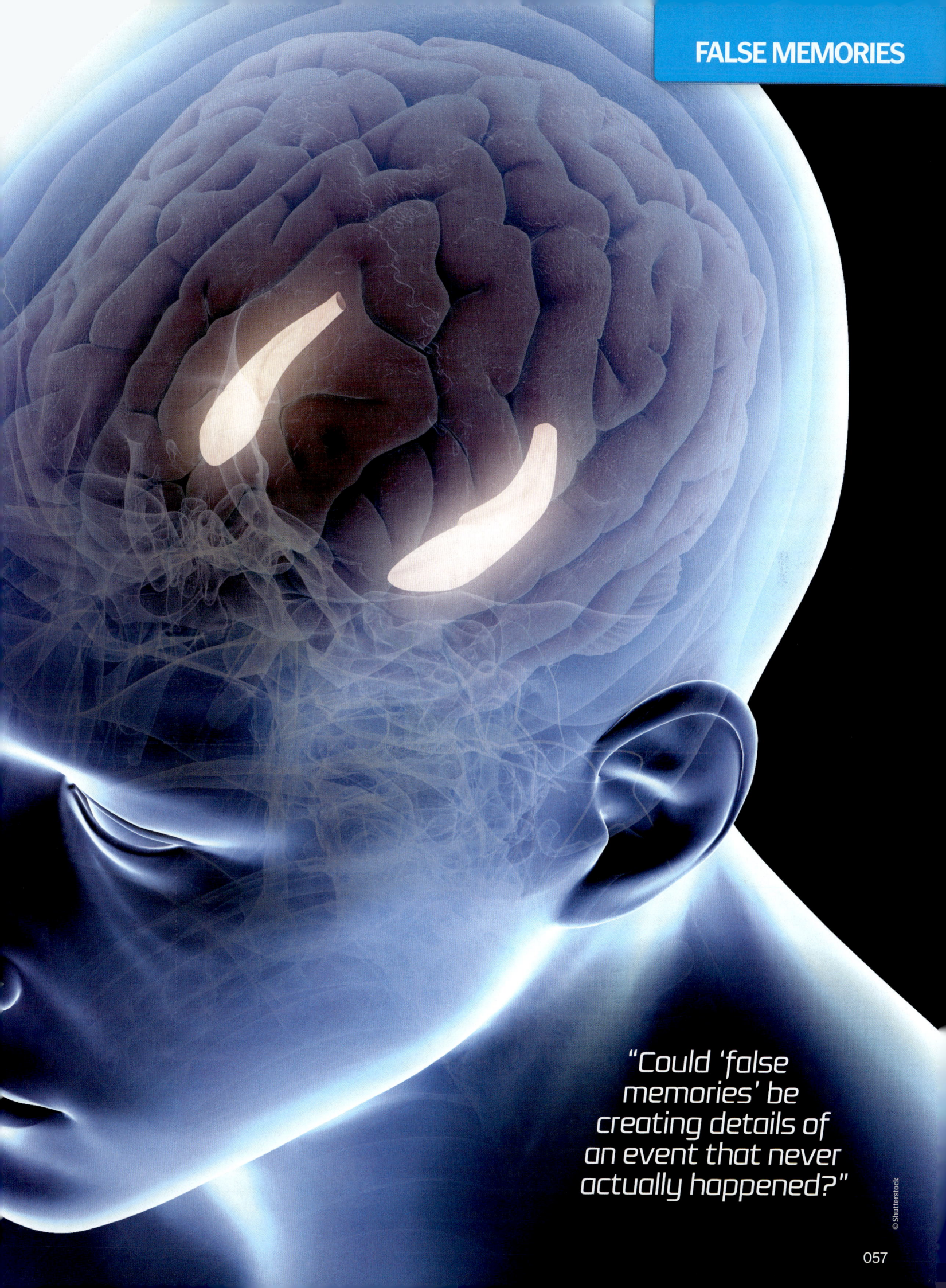

TRICKS AND TRAUMA

Loftus asked test subjects to recall the details of a car crash and found that answers changed depending on the way the questions were asked

THE LOST IN THE MALL TECHNIQUE
WHERE IMAGINATION CAN GENERATE A FALSE MEMORY

You may recall a memory whereby, as a child, you found yourself in a shopping centre or mall alone, having lost a parent among the hustle and bustle of busy shoppers. But what if that memory was a complete fabrication?

Furthering her already extensive research on the topic of false memories, Professor Elizabeth Loftus and her student Jacqueline Pickrell devised an experiment to assess whether or not their study participants could recall a memory of such an event that never occurred. Having consulted with their participants' families, the pair presented test subjects with four short narratives describing events from their childhood, or so it seemed. Unbeknown to the subjects, one of the stories was false and described an event whereby the subject was lost in a mall as a child, around five or six years old. It told of them being left for an extended period of time before being reunited with their parents with the help of an elderly person.

Supported by details unique to each subject's family, the tale seemed perfectly plausible, which was reflected in the results. Around 25 per cent of the participants claimed to be able to remember the event, even going as far as to provide extra information about the event not detailed to them in the experiment. Loftus and Pickrell concluded that the act of imagining the event created a false memory.

Around a quarter of those tested believed that they remembered being lost in a mall despite it never actually happening

> "Our memories can be contaminated and altered in what is now known as the 'misinformation effect'"

strolling past and you recall him wearing a backpack. If you were then asked what colour the backpack was, you might recount seeing red. But then what if self-doubt occurs and, in actuality, you think it was his hoodie that was red? Suggesting an alternative in this way may cause the development of false memory. The truth is that for the majority of us, it's impossible to perfectly recall the events of an entire day, or even the past hour, for that matter.

The truth behind false memories has been readily explored in recent decades, but no more so than by psychologist Professor Elizabeth Loftus. Her work in the mid- to late-1970s revealed how our memories can be contaminated and altered in what is now known as the 'misinformation effect'.

In a series of experiments, Loftus showed a group of participants a video of two cars colliding, then, once they were split into several groups, quizzed the viewers on the events they had just witnessed. One group was asked the straightforward question "How fast were the cars going when they hit each other?", while the other groups were asked the same question, however, the word 'hit' was altered to words such as 'smashed', 'collided' or 'bumped'. Loftus discovered that by altering the way the question was asked the study subjects recalled the event differently. Those asked the question using the word 'smashed' estimated speeds of around 40 miles per hour, whereas those asked using the word 'contacted' estimated around 31 miles per hour. Similarly, when asked "Did you see any broken glass?", the majority of participants correctly answered no, however, those asked the 'smashed' version of the question were more likely to incorrectly say there was broken glass in the video.

What Loftus had uncovered was the effect the verb played in leading the participants to make a false recall of the event. But why is understanding false memories in this way important? The answer lies in how we use our memories. Much more than mental snapshots of our lives, our memories have been legalised and used as evidence to convict so-called criminals since the mid-1970s. However, as DNA evidence became prevalent during the

FALSE MEMORIES

1990s, there was a stark realisation that eyewitness testimony may not be as reliable as was once believed.

At a TED talk in 2013, Loftus recalled a study of 300 cases of individuals where DNA evidence had proven they had been wrongfully convicted, some of whom had served as long as 30 years. In three-quarters of these cases the conviction was a consequence of "faulty eyewitness memory".

There are several factors that can affect eyewitness recall, such as interference, emotion and conflicting information. For example, one common aspect of a crime prone to false memory construction is the presence of a weapon. This was demonstrated by researchers Johnson and Scott in 1976 in an experiment that gave rise to what is now known as the 'weapon focus' effect.

Their experiment centred around the effect of anxiety on the accuracy of testimonies made by eyewitnesses. Participants were asked to sit in a waiting room at a laboratory. A nearby receptionist then excused herself, leaving the test subjects alone. Split into two groups, the first overheard the chatter of the employees discussing equipment failure as an individual walked past the subjects holding a pen in their hands covered in grease.

The second group, placed in the same waiting room scenario, were exposed to a heated exchange, breaking glass and an individual moving past the participants holding a bloody knife. The two groups were then asked to identify each person who had left the laboratory after being shown images of 50 potential candidates.

Surprisingly the members of the first group correctly identified the person walking with the greasy pen 49 per cent of the time, whereas those exposed to the knife-wielding individual only correctly identified them 33 per cent of the time. It's believed that this is due to the higher levels of anxiety felt by those exposed to the knife, with their attention drawn to the knife rather than the person. This suggested a reduction in the accuracy of eyewitness testimonies of people who experience a situation involving a weapon.

Eyewitness accounts of a crime might be examples of false memories due to several influencing factors

STEVE RAMIREZ AND XU LIU
HOW TWO SCIENTISTS SUCCESSFULLY IMPLANTED A FALSE MEMORY

False memories may be unavoidably obtained as a side-effect of the influences and inferences of daily life. However, two MIT scientists, Steve Ramirez and Xu Liu, achieved a work of science fiction back in 2012 when they intentionally implanted a false memory into the brain of a mouse. The experiment centred around provoking a fear memory recall for the mouse at the switch of a laser. Initially placing the mouse in a box, the base of the box delivered a mild foot shock, triggering a fear memory. At the point of memory formation, the pair injected the mouse brain with a light-sensitive light switch protein called channelrhodopsin, which binds to the brain cells involved in memory. After placing the mouse into a second unfamiliar box, a laser was fired at the brain and the light-sensitive channelrhodopsin activated the brain cells and replayed the fear memory to the mouse. This resulted in the mouse's natural fear response of 'freezing', thus proving the scientists had successfully implanted a fear memory into the mouse.

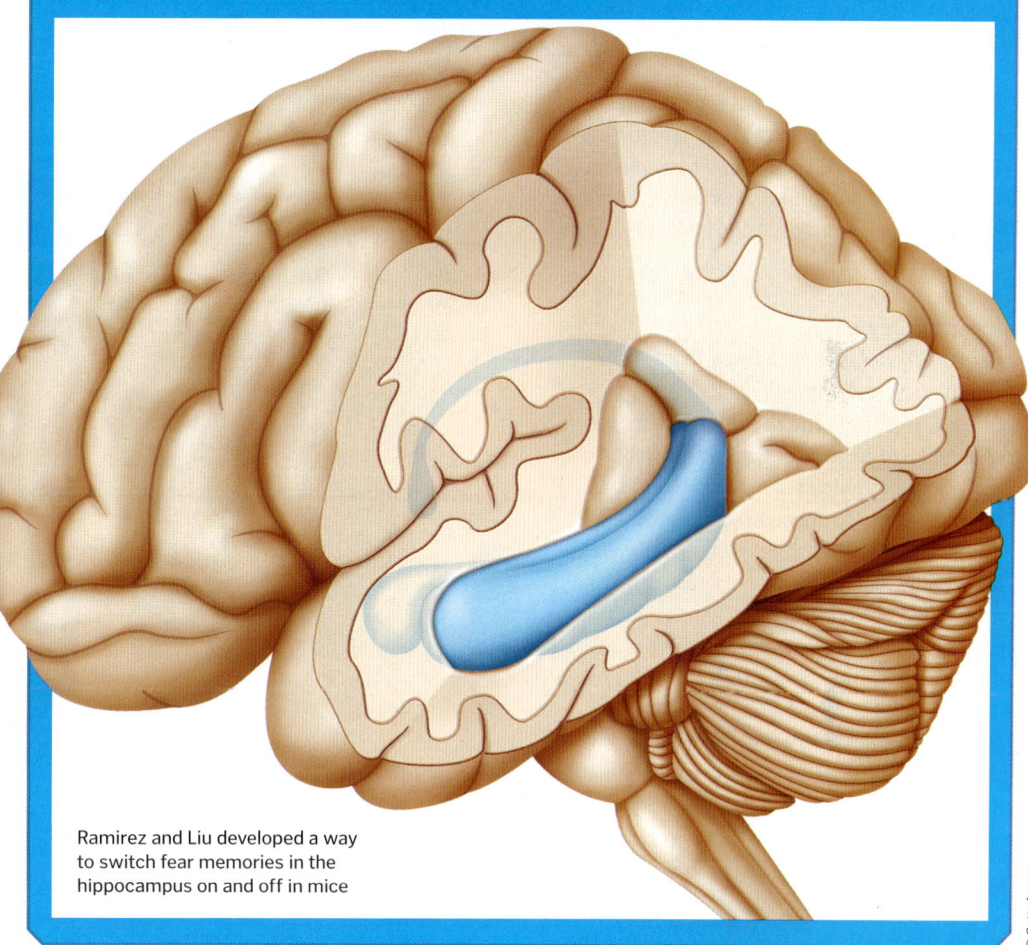

Ramirez and Liu developed a way to switch fear memories in the hippocampus on and off in mice

The presence of weapons, leading verbs or suggestive language aren't the only ways in which false memory can lodge itself in our brains. There are several theories to explain what can lead to the original memory's distortion or alteration.

One such theory is the 'fuzzy-trace theory', outlined by researchers Charles Brainerd and Valerie F. Reyna to explain a memory-creating phenomenon called the Deese-Roediger-McDermott (DRM) paradigm. The DRM paradigm is a false memory task whereby subjects are given a list of related words that they are then required to remember. For example, bed, rest, awake, tired, dream, snooze, nap and slumber. The subjects are then asked to recall the list of words after a certain period of time – different studies have explored a gap anywhere from a few minutes to a few months. What this listicle test found was that subjects typically recalled the word 'sleep', even though it was not on the list, although it does relate to the other words. This is where Brainerd and Reyna's fuzzy-trace theory aims to offer some answers and proposes that there are two types of memory. The first, verbatim, is where we recall the details of a memory vividly and more accurately. The second are gist-like memories, which are imprecise and 'fuzzy' representations of memory, hence the name fuzzy-trace theory.

The research around false memories is still in its relative infancy in terms of fully understanding how our memories are encoded and decoded. Currently, scientists are trying to evaluate how creating false memories might play a role in treating neurological degenerative conditions.

TRICKS AND TRAUMA

EYEWITNESS ACCOUNTS AND RECONSTRUCTED MEMORIES

POLICE LEARNED LONG AGO NOT TO PUT TOTAL FAITH IN AN EYEWITNESS'S MEMORY. DISCOVER HOW THE COGNITIVE INTERVIEW TECHNIQUE CAN HELP BUILD A MORE ACCURATE PICTURE OF A CRIME   Words by **Ben Biggs**

In 1996, Shareef Cousin was convicted of murdering a man during a botched street mugging in New Orleans and sentenced to death. He was exonerated in 1999. Kirk Bloodsworth was convicted of the rape and murder of a nine-year-old girl in Baltimore County, Maryland, in 1985 and sentenced to death. He was exonerated in 1993. Anthony Porter was convicted of a double murder in a Chicago park in 1983 and was only two days from his execution date when a video-taped confession of the real killer led to his exoneration in 1999.

What's the link between these three cases? What made these convictions, for such grave offenses, so shaky? The key witnesses for the prosecution in each, all of whom positively identified the defendants, made eyewitness testimony that was subsequently proved to be completely inaccurate. These witnesses weren't lying about what they saw for personal gain, and nothing or no one had directly influenced their testimony in any way. In the process of recalling the crime they'd witnessed, their brains had simply fabricated parts of the memory that were incomplete or didn't fit a preconceived set of values. A case of joining the dots and then viewing the picture through a lens of personal and cultural bias. To a lesser or greater degree, your brain does this for all your memories in a process that's incredibly subtle and complex.

It can help to understand the way your brain stores information if you imagine your memory works like a computer drive, directly recording video, audio and other sensory information that make up an experience in any given moment. Several regions of the brain are involved in encoding different types of experience and transferring them to other parts of the brain. It enables efficient memory retrieval but does mean these memories are subject to influence. Memories are stored in schemas, units of information that compile patterns of thought and behaviour in relation to people or situations. We don't precisely remember that memory per se; we recall the idea that underpins it in a way that has more meaning for us. So our values, prejudice and past experiences can influence those schemas. British psychologist Sir Frederic Bartlett famously demonstrated how subjective our memories can be in his 1932 experiment, 'War of Ghosts' (see box for more).

To make things more complicated, some of the techniques used to obtain eyewitness testimony can actually end up influencing it. For example, the classic police lineup can be conducted in several ways, including a 'simultaneous lineup' in which the accused isn't included. Witnesses will often finger the person who looks most like their memory of the perpetrator rather than saying that the person they remember isn't in the lineup. They might not even remember the face of the perpetrator clearly because a weapon was present: a gun or knife used in a threatening manner can focus the attention of the eyewitness on the weapon itself to the exclusion of remembering everything else.

A witness might point out the person who looks most like their memory of the perpetrator in a police lineup, even if the actual perpetrator isn't among them

> "Some techniques used to obtain eyewitness testimony can influence it"

Victims of violent crime may have difficulty remembering the perpetrator's face if they used a weapon yet have a detailed recollection of the weapon itself

EYEWITNESS ACCOUNTS

THE WAR OF GHOSTS

In 1932, psychologist Sir Frederic Bartlett conducted an experiment that worked a little like a game of Chinese whispers. He asked each of his subjects to read an obscure piece of Native American folklore they wouldn't have known called the 'War of Ghosts'. He then asked them to recall the story at increasing intervals.

Naturally, the subjects began to forget details from the story as the time between the telling and the remembering increased. As details became hazy, they were adapted or completely removed in the retelling. Bartlett called this 'distortion' and identified three aspects of it. 'Assimilation' made the story more familiar to the subject – Native American details became more British. 'Levelling' made the story shorter as elements were forgotten or deemed unimportant and removed. 'Sharpening' changed the order of the story and added certain details. Bartlett gained wide recognition for his work, as the experiment showed how a person's memory can be heavily influenced by their own knowledge and experiences.

The cognitive interview technique was devised in 1992 to help form a more accurate eyewitness account of a crime. The interviewer asks the witness to recall the events surrounding the crime in several ways, including remembering the sequence of events in a different order, and remembering them from the perspective of the criminal or another witness. It's thought that, by changing the order of the narrative and perspective, the witness's reliance on prior knowledge to retrieve the memory is reduced. This technique can also help jog their memory of other details they might not have otherwise remembered via a conventional interview method.

The cognitive interview technique was used to verify information and obtain a confession from Karen Matthews in the 2008 case of kidnapping victim Shannon Matthews (her daughter)

THE COGNITIVE INTERVIEW TECHNIQUE
DETECTIVES CAN DEPLOY THIS PSYCHOLOGICAL INTERVIEW TECHNIQUE IF THEY THINK IT WILL IMPROVE THE ACCURACY OF THEIR EYEWITNESS' ACCOUNT

Putting the crime in context
The detective asks the witness to recall the events surrounding the crime and the crime itself using all five senses. If the witness cannot clearly recall what they saw, for example, then a clear memory of what they heard might help build the picture.

Changing perspective
The witness is asked to imagine the scene from someone else's perspective: another witness standing across the road, the victim, or even the perpetrators of the crime.

Turning back time
We tend to remember the most recent events more clearly, so the detective asks the witness to try to recount the events of the crime in reverse order.

Devil in the detail
Finally, the witness dwells on the incidental, seemingly inconsequential details of the crime, such as the type of shoes the criminal was wearing, or the number plate of the car that was approaching the scene, or the colour the traffic lights were at a precise moment.

061

TRICKS AND TRAUMA

THE MANDELA EFFECT

FALSE MEMORIES ARE OFTEN SHARED GLOBALLY, RAISING QUESTIONS ABOUT COLLECTIVE HUMAN MEMORY

Words by **Nikole Robinson**

Memory is a tricky thing. With so much happening all around us, there's no way to take in and remember everything we experience, and sometimes we don't recall details of what we do remember correctly. We can also have a completely different interpretation of events than someone else who experienced the same thing. This can happen on a global scale and often with extreme differentiations. To the people holding these false memories, their individual impressions feel so real to them that they are unable to accept that their memory is wrong, even when there is concrete evidence that proves otherwise, often expressing shock, confusion and sometimes even anger.

Many of the most widespread false memories are rooted in pop culture or are of events that are well known around the world. Though some may seem insignificant or are simply a minor detail – or, naysayers argue, can be put down to basic human error – it is still interesting to see how individual interpretations vary, especially when some inaccuracies are so strange or easy to disprove but are shared by thousands.

Some of the more wild theories suggest that parallel universes could be at play, with a multiverse full of unlimited possibilities creating chances for every version of events. However, more acceptable underlying contributors to this false memory phenomenon could be misinformation, the brain linking events to associated information that alters details of the memory or the human quality of being inclined to accept the suggestions of other people. The spread of misinformation – especially in the era of the internet – may also be an underlying cause.

NELSON MANDELA'S 'DEATH'

The most famous false memory became a moniker for the phenomenon when it was revealed that thousands of individuals could remember Nelson Mandela passing away in the 1980s during his time in prison. In fact, he didn't die until 2013.

THE MANDELA EFFECT

THE BERENSTAIN BEARS

A popular children's literature franchise, many who grew up reading the series remember the name spelt as Berenstein. One attributing factor to this false memory may be unofficial knock-off merchandise with the name spelt incorrectly.

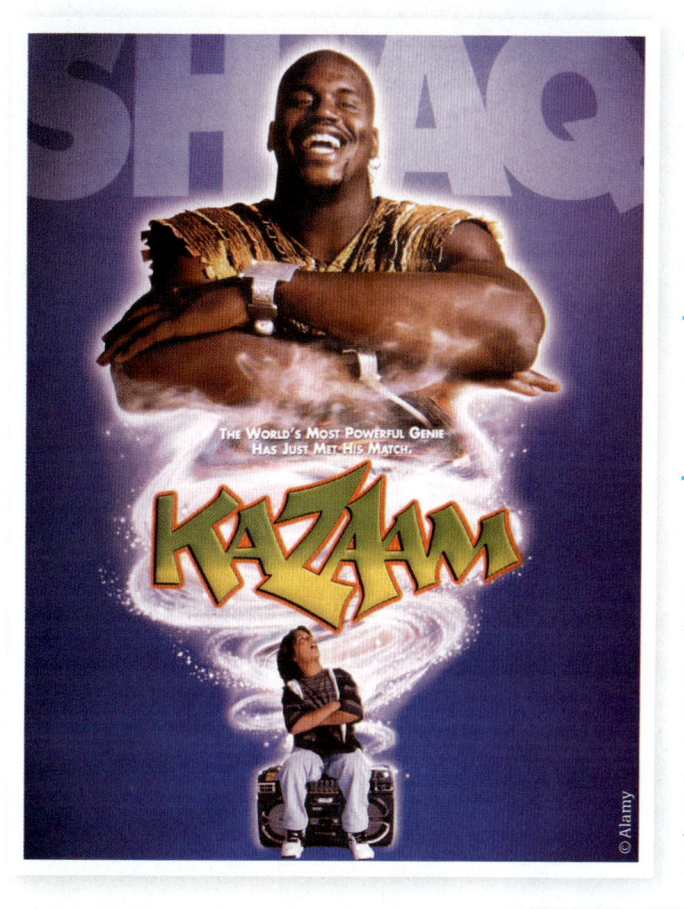

SHAZAM

"Many remember the show being called 'Looney Toons', not the correct Looney Tunes"

Many people can remember a 1990s movie that starred stand-up comedian Sinbad as a genie, but no such film exists. People could be mistaking it for the 1996 film *Kazaam* starring NBA All-Star Shaq.

BOLOGNA STATION CLOCK

The clock at Bologna Centrale Station was damaged in the Bologna bombing in 1980. A 2010 study showed that 92 per cent of its respondents believed that the clock had been stuck at the same time since, although in reality it was fixed soon after the attack.

OUT OF TUNE

Possibly because of the 'oo' in 'Looney' or because of its connotation with the word 'cartoon', many who enjoyed the adventures of Bugs, Daffy, Sylvester and Tweety remember the show being called 'Looney Toons', not the correct *Looney Tunes*. The name was inspired by Disney's *Silly Symphonies*.

THE DEESE-ROEDIGER-MCDERMOTT PARADIGM

A procedure to investigate false memory in humans, subjects are presented with a list of words relating to a keyword that is not present in the list – 'peach', 'apple' and 'banana' might be present, but not the word 'fruit', which links them. There are also a few words in the list that are unrelated to this keyword as a distraction. Participants are then asked to recall as many words as they can remember. If one of these keywords is recalled by the subject, this indicates a basic mistake in their memory processors. This test shows us that in some people, unconscious associations could be behind the altering of memories, causing them to experience a false memory. Further studies using the paradigm have shown that false-memory retrieval reveals different neural activity in the brain to the recovery of a genuine memory.

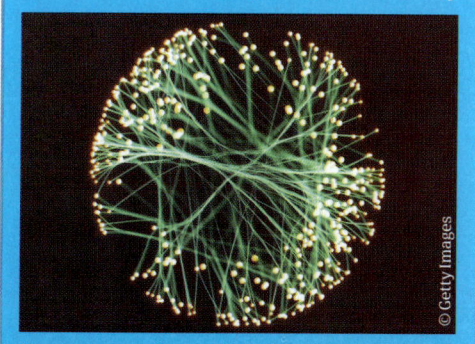

Neurons carry electrical impulses in our brains

TRICKS AND TRAUMA

SEX AND THE CITY

A lot of people are absolutely certain that they remember the show – which ran between 1998 and 2004 – being called 'Sex in the City', but it's not. Some believers have even found merchandise suggesting 'in' is correct, but these could be unofficial goods or misprints. The show, which was created by Darren Star and inspired two films, has always been called *Sex and the City*.

MONOPOLY MONOCLE

Though it would pair nicely with his top hat and cane, the Monopoly man never wore a monocle. However, many people are surprised when they pull out the game and see that it isn't there, swearing they remember him having one.

WE ARE THE CHAMPIONS...

...of the world! If you belt that last part out at the end of Queen's famous hit, you'll be surprised to find out that Freddie Mercury doesn't sing it along with you like in the previous verses.

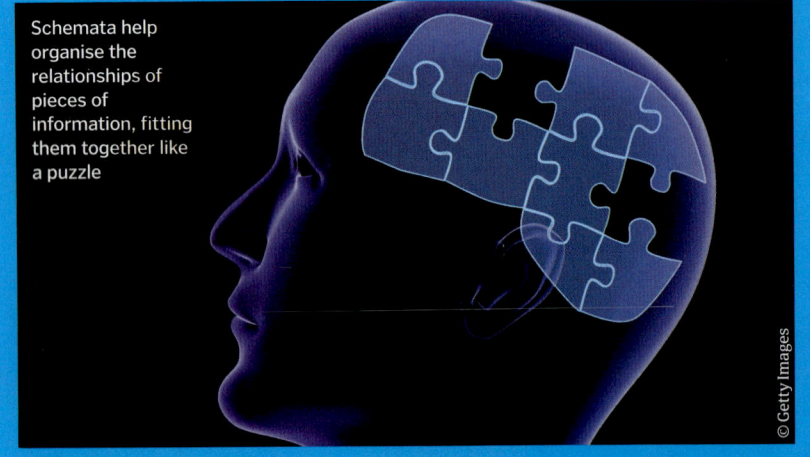

SCHEMA-DRIVEN ERRORS

It could just be the fact that we're remembering what is typically expected that accounts for these errors in memory. Schemata are patterns in thought that organise information and the relationships between different things. They provide a framework for future understanding and therefore influence the absorption of new material into our brains. People are more likely to notice things that fit into their established schema, while details that don't are reinterpreted or distorted to fit the pattern. As an example, when participants in a study were asked to recall the contents of a psychologist's office, they tended to remember consistent and commonplace items such as bookshelves and desks and omit any inconsistencies they felt didn't typically belong. People's individual schemata have the tendency to remain unchanged even when contradictory information is presented. While schemata are useful in helping us process and understand new information, they also play a part in distorting memories.

Schemata help organise the relationships of pieces of information, fitting them together like a puzzle

THE MANDELA EFFECT

HAVE A BREAK
Some remember these chocolate-covered wafer biscuits having a hyphen in the name, recalling enjoying a tasty Kit-Kat. They're often frustrated or confused when they find out that it isn't spelt with one.

CURIOUS GEORGE
The titular character of the *Curious George* books and TV series has been around since 1939. People across generations remember him having a tail – after all, a monkey usually does – but George gets along fine without one.

FEBREZE
You might even use this cleaning product every day at home and still think it's always been spelt 'Febreeze', looking at the bottle in disbelief. Don't worry, you're definitely not the only one to remember it being spelt that way.

"People remember George having a tail – a monkey usually does – but he gets along fine without one"

"LUKE, I AM YOUR FATHER"
When Luke Skywalker confronts nemesis Darth Vader in Cloud City, Vader reveals to Luke one of the biggest surprises in film history. Yet despite its memorability, many misphrase the quote. Vader actually says, "No. I am your father."

TRICKS AND TRAUMA

SKETCHY MEMORY

The name of the footwear company is often erroneously thought to be spelt Sketchers, when it is in fact Skechers. This may be because it looks more natural to us in English or because of mispronunciation.

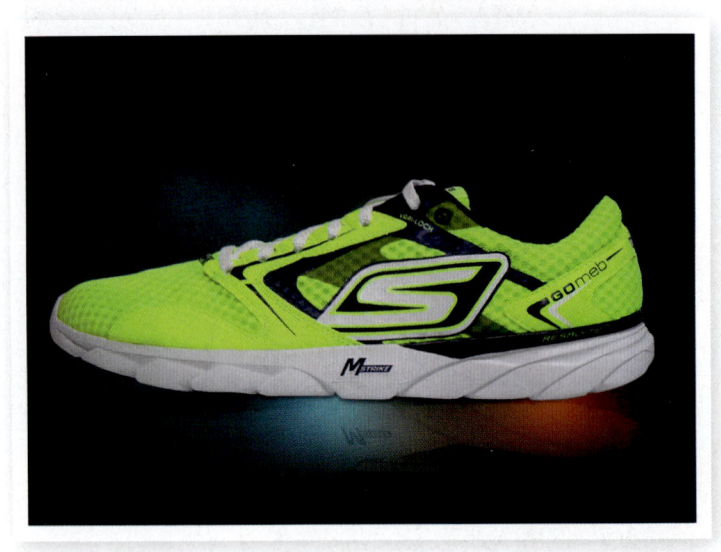

STATE OF CONFUSION

There are 50 states that make up the United States of America, but large groups of people are adamant that there are 51 or 52. They could be thinking of Puerto Rico, which is a territory, or D.C., which is a district.

CHANGING CHARTREUSE

What type of colour is chartreuse? If your mind went straight to a shade of pink, you may be one of those experiencing a false memory. It's actually a yellow-green, named for its similarity in colour to a French liqueur.

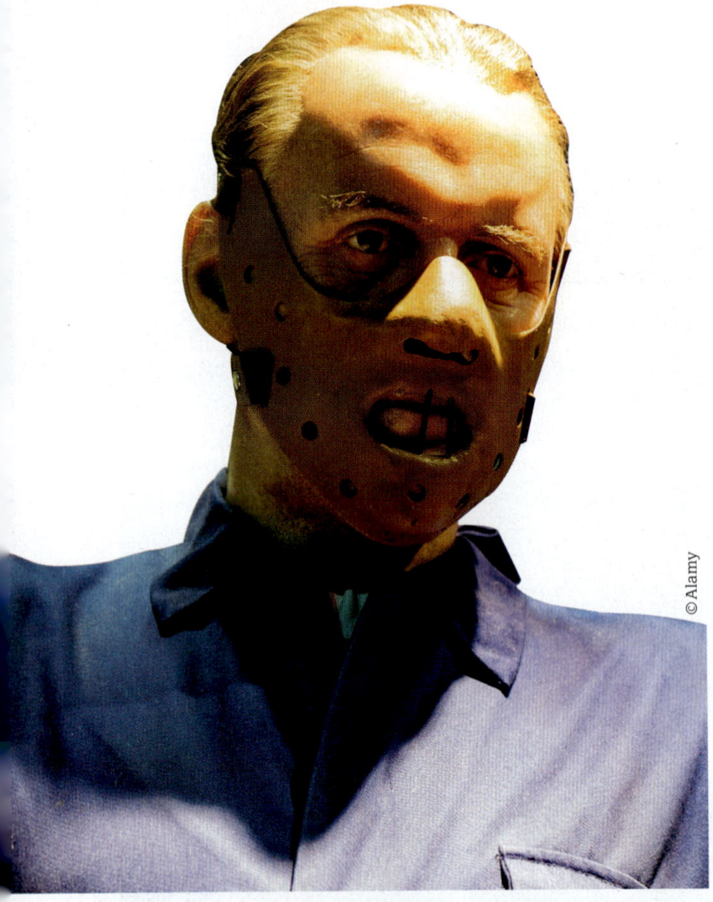

HELLO, CLARICE

In the 1991 psychological horror film *The Silence of the Lambs*, FBI trainee Clarice Starling visits Hannibal Lecter for advice on a case. Many remember him eerily greeting her, but in reality it was a simple "Good morning." Not quite as chilling.

MOTHER TERESA

Roman Catholic nun Mother Teresa was canonised as a saint by Pope Francis on 4 September 2016. Many were surprised at the news because they recall this happening as far back as the 1990s under Pope John Paul II. This may be the result of people mixing up the many recognitions Mother Teresa did receive while she was alive.

THE MANDELA EFFECT

LOST SRI LANKA

If you were asked to point out Sri Lanka on a map, you'd be correct in remembering it's an island near India, but in which direction does it lie off the coast? It's to the southeast, but it's a common mistake that people look straight to the south.

NO CORNUCOPIA

A well-known manufacturer of printable T-shirts and other items of clothing, Fruit of the Loom's logo features bunches of grapes and a bright, red apple. However, many are confident that there used to be a cornucopia cupping the fruit.

THE PARALLEL UNIVERSE THEORY

One explanation for this phenomenon that's a little more out of this world is born from quantum mechanics and theorises that people on the false side of a memory may have somehow crossed from a parallel universe – where their version holds true – into ours. In multiverse theory, it's hypothesised that there are an infinite number of parallel universes alongside our own, each with its own laws of physics and individual timeline. Although it's said to be impossible for these separate universes to interact – also meaning that multiverse theory is impossible to prove or disprove – fluctuations in space-time or a collision between multiverse 'bubbles' could cause people from other realities to have switched places with a doppelganger in our own, bringing with them their memories of a similar but incorrect version of events. It sounds like science fiction, but many high-profile astrophysicists – including the late Stephen Hawking – support the multiverse theory.

Could the existence of countless alternative universes explain this bizarre phenomenon?

HENRY'S TURKEY LEG

King Henry VIII ruled England from 1509 until his death in 1547, living a decadent life of debauchery. Many mistakenly remember a version of his portrait where the portly king holds a turkey leg, but no such portrait has been found to exist.

TIANANMEN SQUARE TANK MAN

In an iconic image taken the day after the Chinese military suppressed the 1989 Tiananmen Square protests by force, an unknown protestor blocks the path of four tanks. The final fate of the man is unknown, but he wasn't run down by the tanks as some believe happened.

067

TRICKS AND TRAUMA

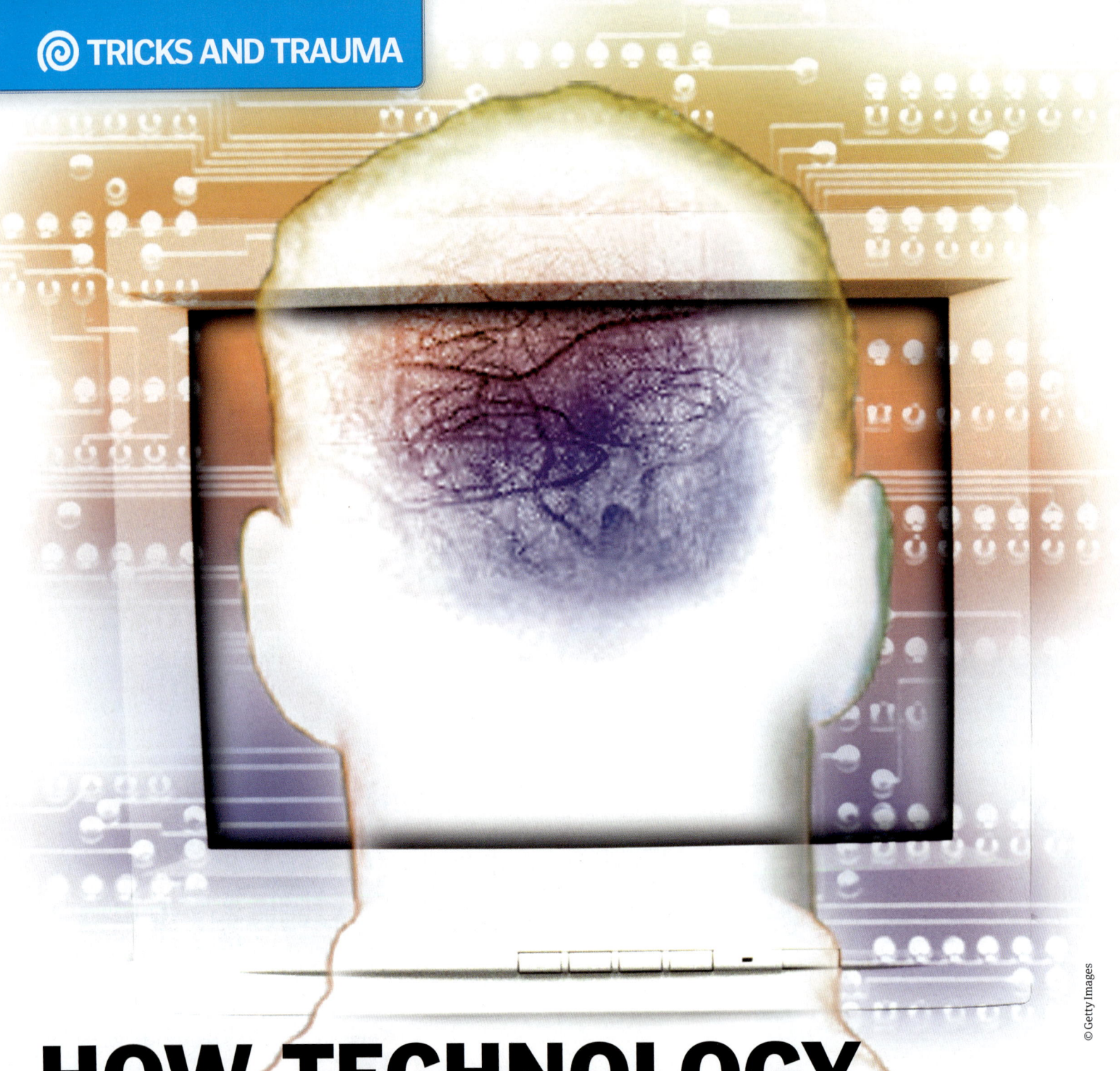

HOW TECHNOLOGY CHANGES OUR MEMORY

HAS OUR ABILITY TO RECALL EVENTS, STATS AND FACTS WANED BECAUSE OF A RELIANCE ON SMARTPHONES AND COMPUTERS? Words by **Aiden Dalby**

Parents often tell their children that looking at smartphones and computers all day will rot their brain, but is this an old wives' tale or are our gadgets really having an effect on our brains? The majority of us spend a significant amount of our day with technology either for work or just to pass time, but over the past ten years the amount of time spent on these devices has grown substantially. Those who grew up without smartphones will recall when they could recite phone numbers from memory, but the same cannot be said today. Is technology to blame? Are parts of our brains that were once active now dormant thanks to our reliance on gadgets?

The short answer is no. Technology hasn't ruined our memory; instead it could be changing how our memory is being used. It is believed that a different part of our brain is active when we use these devices to look up information. In 2011, researchers found that when we look up facts and statistics online, we are less likely to remember that content, but our brain is still memorising

"Are parts of our brains that were once active now dormant thanks to gadgets?"

TECHNOLOGY'S IMPACT

Younger smartphone users are likely to check their device more regularly

Switching from one device to another prevents the brain from focusing

Those who use their smartphones during concerts will have weaker recollection compared to those who don't

information. This is known as the 'Google effect'. Our brains are not retaining the information we have looked up but instead they are remembering where the information can be found. So our brains are still working to remember, but we may not be memorising the information that we want to. Going back to the example of remembering phone numbers, before smartphones we would manually enter phone numbers and doing this repeatedly would allow it to be ingrained into our memory. But because we can simply call someone with shortcuts on our smartphones now, there is no need for our brain to remember the number; instead we remember how we can find it.

Another experiment sent groups of people out on a tour, with some participants encouraged to use their smartphones to take photos and videos of the experience while the others were told not to. The study found those who took photos and posted them online had a poorer recollection of the events compared to those who didn't. It is ironic that in our attempt to preserve a memory by capturing footage of it we could be preventing that very thing from happening.

Another way technology can affect our memory is by being a distraction. Our devices offer a seemingly endless amount of content from games to videos and social media. There is always something new to look at, and if the brain is unable to focus then its ability to memorise what is happening is going to be impaired. During tasks when we should be focusing, such as studying for an exam, the temptation to turn to our phone or open a new tab on an internet browser to see what's going on can be too great to resist, but constantly switching from task to task prevents the brain from retaining information. A 2015 survey of over 15,000 adults in the USA found that 81 per cent of people keep their smartphones close to them at all times and around half of the participants checked their smartphone a few times every hour or more.

The survey also found that one in five 'young Americans' checked their phone every few minutes. This constant switching from one thing to another has been shown to prevent long-term memories from forming.

There are other ways in which technology could be adversely affecting our ability to remember. Looking at smartphone screens and computer monitors for long periods of time can affect the quality of sleep that we get. Our brains need a good night's rest in order to allow them to process the information taken in throughout the day and turn short-term memories into long-term ones.

The long-term affects of technology's impact on our memory are unclear as the amount of research available today is limited, but we do know of some of the affects it is already having. Too much time spent on our phones and computers can prevent us from forming lasting memories, but by being more mindful of how much we use them we might be able to change this. As with most things, moderation is key.

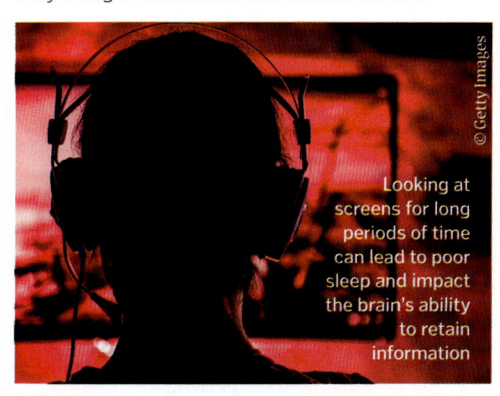

Looking at screens for long periods of time can lead to poor sleep and impact the brain's ability to retain information

MEMORY-SAVING TECH

A study in 2013 found that playing brain training games for 15 minutes a day, five days a week can improve working memory. Elevate is one of the most popular apps on the market; available for both iOS and Android devices, it has won both Apple's App of the Year and Google' Editor's choice awards. It's free to download, and you can select what you want your goal to be (improved focus, processing information, retain more of what you read etc.) and Elevate will tailor the training programme to you. You are given daily brain workouts of five games out of more than 35 to complete, and if for any reason you forget to complete them the app will send a notification reminding you of how many you have left. When you complete games the difficulty will change depending on your results to help you improve.

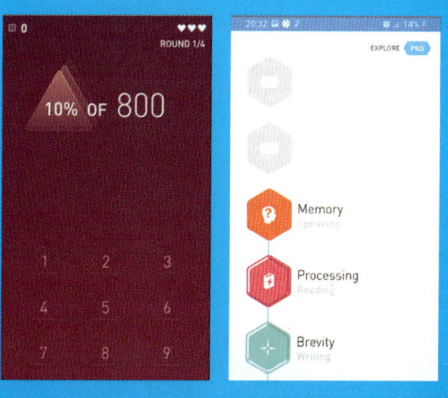

Elevate works on improving several parts of the brain, including memory

069

TRICKS AND TRAUMA

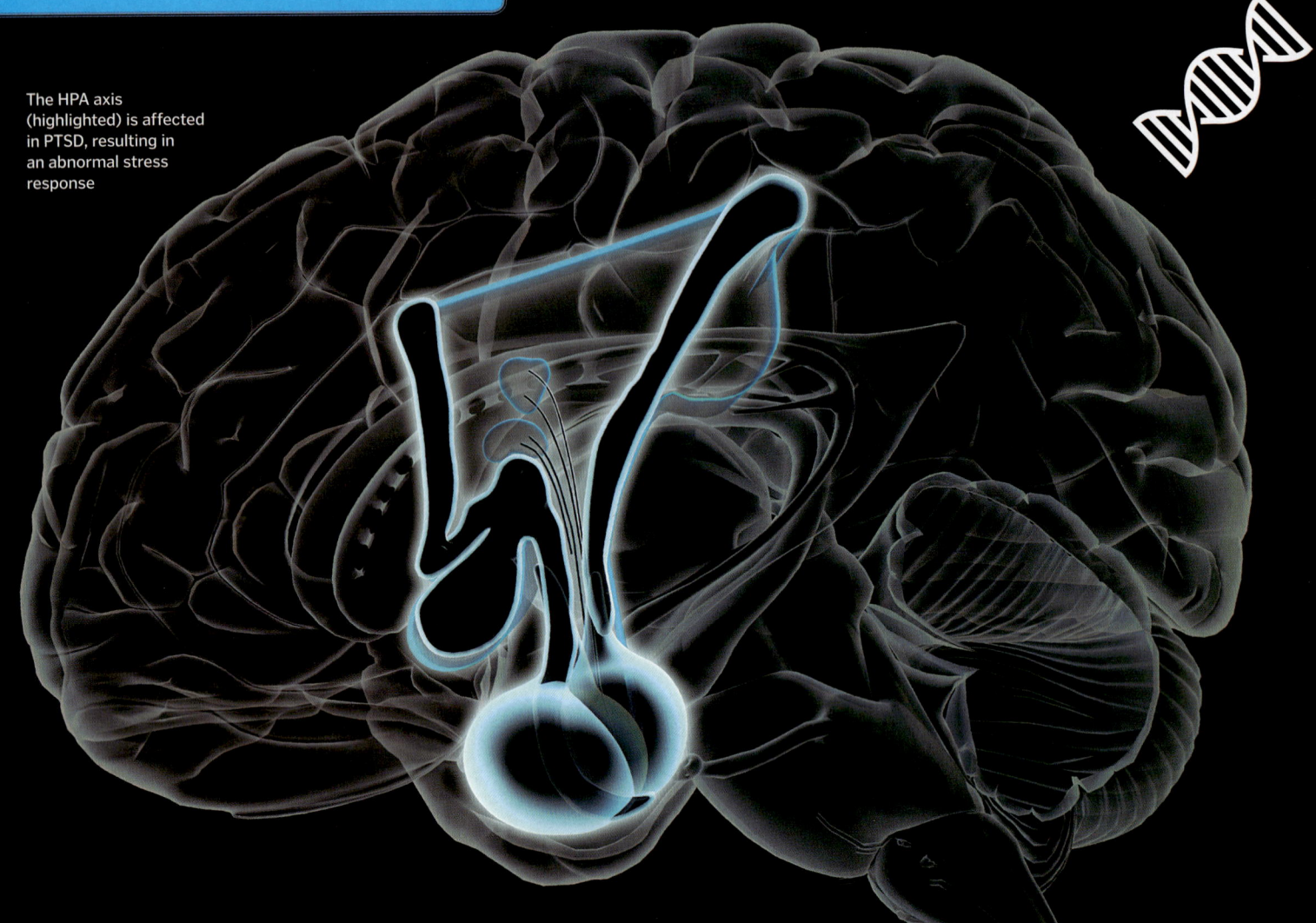

The HPA axis (highlighted) is affected in PTSD, resulting in an abnormal stress response

TRAUMA
& MEMORY LOSS

TRAUMA CAN DESCRIBE EITHER A PHYSICAL INJURY OR A DISTRESSING EXPERIENCE. ALL TYPES OF TRAUMA CAN LEAD TO POST-TRAUMATIC STRESS DISORDER, A SYMPTOM OF WHICH IS MEMORY LOSS

Words by **Josie Clarkson**

Post-traumatic stress disorder (PTSD) refers to the set of negative psychological symptoms that follow exposure to actual or threatened death, as defined by the latest *Diagnostic and Statistical Manual of Mental Disorders* (DSM-V), which is used to diagnose mental conditions. The kind of events that go on to trigger PTSD can be anything traumatic to the individual, including wars, serious accidents and physical or sexual abuse. The psychological aftermath of trauma varies between patients, but the main symptoms revolve around anxiety and stress. Another symptom listed in the DSM-V is an "inability to remember an important aspect of the traumatic event". This memory lapse is known as 'dissociative amnesia' and is not caused by any sustained injury or drugs.

TRAUMA AND MEMORY LOSS

Amnesia is a partial or total loss of memory; dissociative amnesia is where the traumatised individual blocks out information about the traumatic event – dissociating themselves from it – leaving gaps in their memory of the time period spanning the event. Some psychologists believe that these memories have not been wiped but have instead been buried in the subconscious, unavailable for conscious recall but able to resurface when triggered by things related to the trauma. This theory also explains why people with PTSD experience vivid flashbacks of the event, which involve a sudden, involuntary reliving of the traumatic experience. Therefore, these flashbacks are the suppressed memories being forced into the person's consciousness.

Memory is a complex process that is broken down into various categories depending on what is being remembered. Declarative memory refers to the memory of facts and events and is partially affected by PTSD. In patients with PTSD, this declarative memory loss can extend beyond the dissociative amnesia of the traumatic event and cause problems remembering ordinary, unemotional facts and events. Declarative memory can be assessed with visual memory tests and verbal memory tests, the difference being the format (i.e. pictures or words) in which the items are presented to the participants. Through people with PTSD taking these memory tests, researchers have found that PTSD is linked with a subtle impairment in verbal declarative memory, with visual declarative memory being less affected. The learning process is more disrupted than the process of storing and retaining memories, meaning the main problem is forming new memories. The severity of these memory difficulties is directly related to how long a person has had PTSD for, so the memory of people with a long history of PTSD is most impaired.

Scans of PTSD patients' brains have revealed abnormalities in a brain area called the hippocampus (responsible for acquiring new memories) and the hypothalamic-pituitary-adrenal (HPA) axis. The HPA axis represents the communication between the hypothalamus, the pituitary glands and the adrenal glands, which influences behaviours such as mood, temperature control and the immune system through the release of hormones. These brain systems are involved in the stress response, so abnormalities here indicate a disrupted stress response.

A hormone called cortisol is released by the adrenal glands when people and animals are stressed. Researchers studying the stress

> "The memory of people with a long history of PTSD is most impaired"

Domestic abuse can cause PTSD in those who witness or experience it

An increased startled response is a symptom of PTSD

RECOGNISE PTSD SYMPTOMS

These symptoms are paraphrased from the DSM-V and must have been occurring for at least a month in individuals aged six years or above for a diagnosis of PTSD to be given.

- Intrusive distressing memories of the trauma.
- Nightmares related to the traumatic incident and disturbed sleep.
- Avoidance of stimuli associated with the event.
- Dissociative amnesia.
- Easily startled – also known as hypervigilance.
- Uncontrollable negative emotions – e.g. anger / guilt / sadness / stress.
- Apathy and disinterestedness in activities.
- Self-destructive behaviour.
- Anxiety.
- Physiological symptoms – e.g. sweating and raised heart rate.

TRICKS AND TRAUMA

The hippocampus (highlighted in red) in people with PTSD is smaller than in people not suffering PTSD

DEPRESSION, PTSD AND MEMORY

Almost half of people diagnosed with PTSD also have a diagnosis of depression. The two mental disorders are so closely linked that it has been suggested that PTSD with depression is its own subtype of PTSD. This is perhaps unsurprising because they share symptoms, such as sleep disruption, trouble feeling pleasure and, crucially, memory difficulties. These memory difficulties are very similar to those seen in PTSD – depressed patients perform poorly in verbal memory tasks, but – unlike people with PTSD – in visual memory tests as well. They also share certain brain abnormalities, such as overactive HPA axes, resulting in more cortisol being released and smaller hippocampi. Furthermore, both depression and PTSD have been proposed as risk factors in increasing a person's likelihood of developing dementia, a collection of diseases of progressive memory loss.

There is a large body of research linking depression with dementia, but the link with PTSD has only recently emerged from a couple of pioneering studies with veterans. Scientists investigating this link have suggested hormonal imbalance, HPA dysfunction and loss of hippocampal volume could be to blame, but more investigative research needs to be done before a causal relationship between PTSD and dementia can be identified.

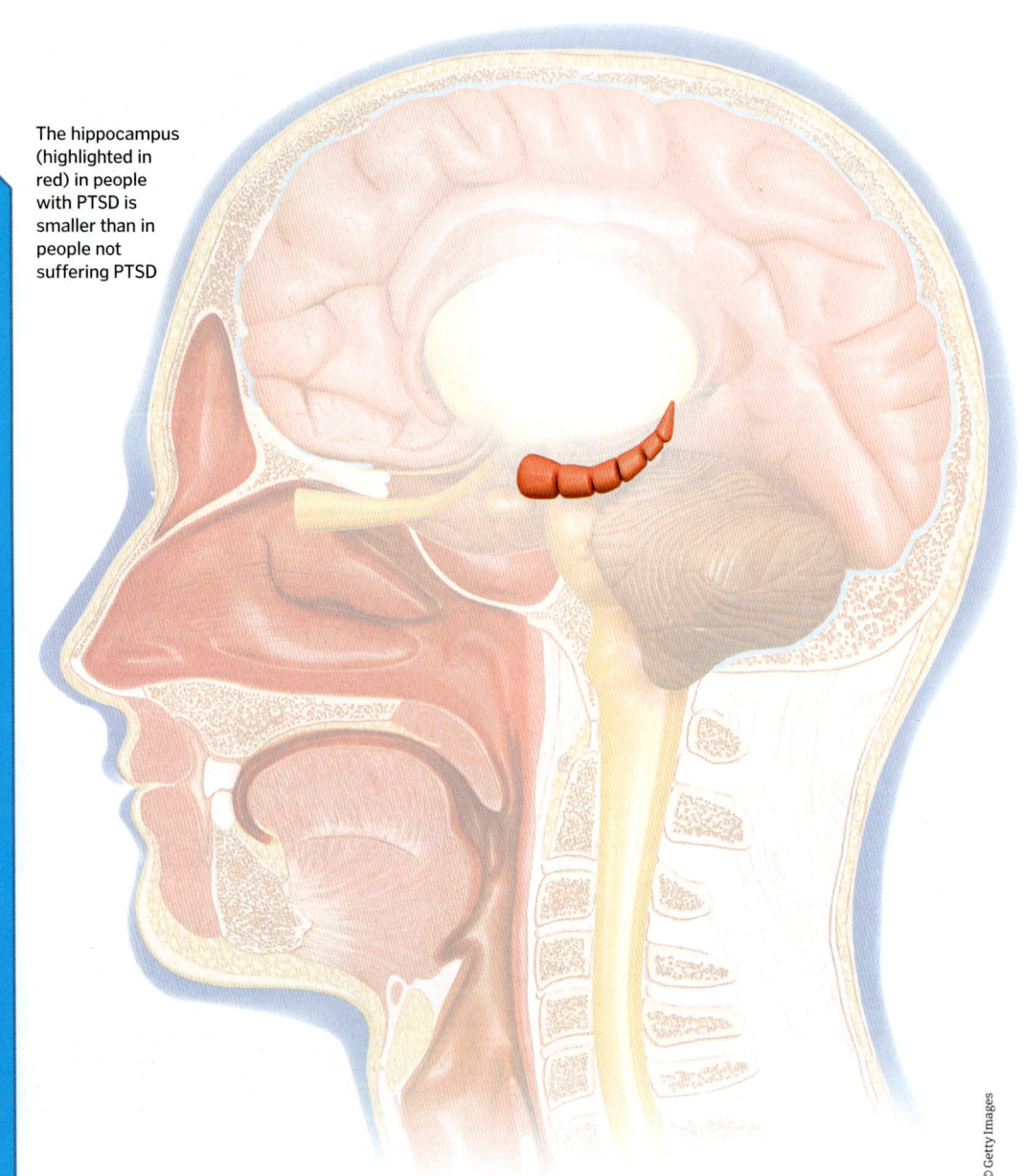

"People with a poorer memory may be more susceptible to PTSD"

response of animals have found that cortisol is toxic to the hippocampus, so prolonged periods of stress, like in PTSD, can damage the hippocampus. During stress, cortisol is also released onto the prefrontal cortex, an area at the front of the brain that contributes to executive brain functions like working memory, behaviour and emotion. Interestingly, both the volume and activity of the prefrontal cortices of PTSD patients are lower than those of people without PTSD. These reductions are especially prominent in children.

The DSM-V diagnosis is only valid for people aged six and above. Even then, the brains of children over age six diagnosed with PTSD are affected differently to adults with PTSD. They still exhibit hippocampal abnormalities but, instead of reduced volume, they have the same volume as age-matched children without PTSD, just less brain cell activity there. However, as these traumatised children grow up with continual stress, their hippocampi are likely to reduce in volume to mirror the abnormal volume seen in adults with PTSD.

Another way to look at the relationship between trauma and PTSD is that pre-existing memory and learning problems could be risk factors for developing PTSD. That is, people with a poorer memory – and, therefore, some issues with hippocampal functioning – may be more susceptible to PTSD if they experience trauma. It's a kind of chicken and egg scenario – which came first, the memory deficits or the PTSD?

There is some evidence from studies of identical twins that memory problems do precede PTSD. In the twins studied, one was a war veteran with subsequent PTSD and the other was neither a veteran nor had PTSD. Both twins in the pairs had smaller hippocampi and poorer verbal memory ability compared to the average person. Therefore, some memory problems were present in both twins before one of them developed PTSD.

Twins are a useful way to determine whether a condition is caused by biology or the environment as they have identical genes. Consequently, this research supports memory impairment being a risk factor for PTSD rather than PTSD causing memory deficits.

Neuroscientists Jennifer J. Vasterling (professor of psychiatry at Boston University School of Medicine) and Kevin Brailey (staff psychologist for the Center of Returning Veterans) theorised a 'downward spiral' where both PTSD and memory

TRAUMA AND MEMORY LOSS

problems affect each other. Their theory accepts that having poorer memory and learning skills makes someone more likely to develop PTSD and suggests this is because they have fewer psychological resources to cope with trauma. However, they also argue that the onset of PTSD further impairs cognitive functions such as memory and learning. This cognitive disruption not only interrupts daily tasks but can also interfere with psychological treatment.

Ultimately, the distressing nature of PTSD makes it difficult to study. The dissociative amnesia presents a hurdle if the subject struggles to remember the traumatic event. Plus, participating in a study centred around their PTSD could trigger flashbacks and other traumatic symptoms, so many patients may be reluctant to take part. Understandably, ethical guidelines prevent researchers inflicting trauma on subjects in order to study it, so they must rely on the few PTSD patients who are willing to be studied. Therefore, reviews of multiple studies are crucial to make the most of the research that does exist and develop treatment strategies that do not rely on memory.

People with PTSD may experience nightmares and have difficulty sleeping

Verbal abuse by parents can cause PTSD in children

Soldiers often return from combat with PTSD due to the trauma they witnessed

TRICKS AND TRAUMA

THE CURIOUS CASES OF HENRY AND EUGENE

MEET THE TWO PATIENTS WHO REVOLUTIONISED OUR UNDERSTANDING OF MEMORY AND HABITS

Words by Josie Clarkson

Neuroscience research relies on patients with existing brain damage to provide an insight into the workings of the brain. This reliance is largely due to ethical guidelines understandably preventing scientists from manipulating live human brains. However, guidelines haven't always been so strict, and lobotomies – removing chunks of the front of patients' brains to 'treat' mental health conditions – were rife until as recently as the mid-20th century. A similar procedure termed a lobectomy also exists, which is the removal of a lobe of a diseased organ. These used to be performed on the brain but are now usually restricted to the lungs. A misguided brain lobectomy removed the medial temporal lobe of a patient known as H.M., who was later revealed to be a young man by the name of Henry Molaison.

Brain damage can also occur naturally through viral infections such as encephalitis, where a virus spreads from the body into the brain, causing inflammation and the destruction of brain cells. This happened to patient E.P. (Eugene Pauly) and resulted in the loss of his medial temporal lobe.

The medial temporal lobe is a collection of brain structures located in the centre of the brain. It includes the hippocampus, entorhinal and perirhinal cortices, amygdala and subicular complex, which work together to process new information and combine it into memories. Henry was diagnosed with a form of epilepsy occurring in the temporal lobe after being hit by a bike at the age of seven. His condition was managed by a cocktail of anti-epileptic medications that allowed him to live a normal life at his home in Connecticut, USA. However, his epilepsy gradually worsened throughout adolescence until, at the age of 27, his

"Henry would forget he'd eaten his bacon roll in the morning and repeat the task, resulting in him eating up to four for breakfast"

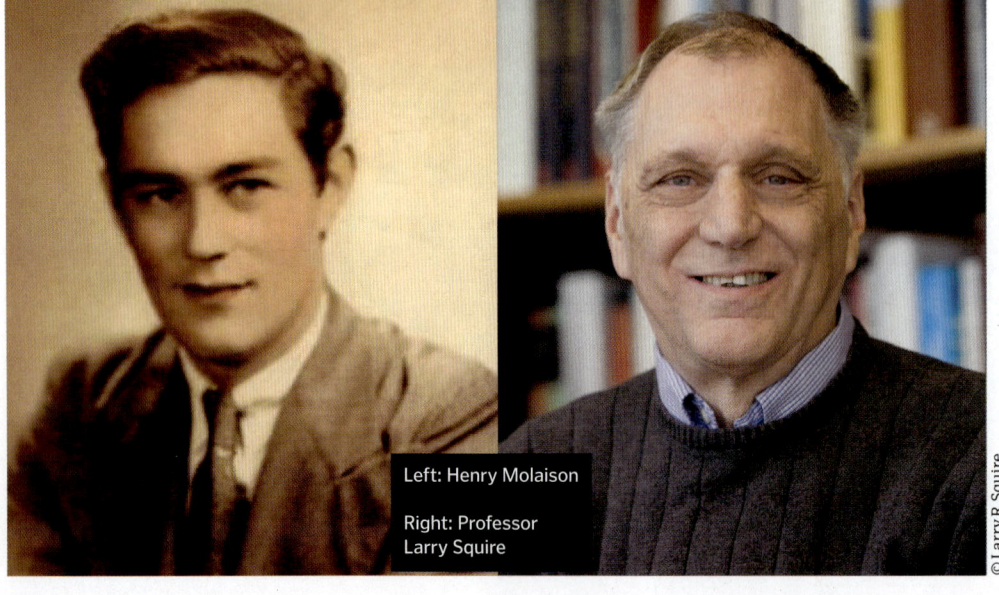

Left: Henry Molaison

Right: Professor Larry Squire

THE CURIOUS CASES OF HENRY AND EUGENE

TRICKS AND TRAUMA

seizures were so disruptive that he was forced to give up his work on an assembly line. It was then, in September 1953, that Henry sought the drastic medical intervention that was to change his life – and neuroscience – forever.

Without the complex tools we have today, such as MRI scanners, which allow doctors to peer inside the body, Henry's surgeon, Dr Scoville, removed a 54.5mm-long area on the left side of his brain and a 44mm-long portion on the right: his medial temporal lobe.

Without his medial temporal lobe, Henry's epilepsy was partially alleviated, but the operation was not wholly successful, as he still required anti-epileptic medication. He was also left with severe amnesia, a form of memory loss that can present in different ways.

For Henry, this meant he was unable to form new memories after the operation (known as retrograde amnesia) and had forgotten memories he had formed in the couple of years immediately before his operation (partial anterograde amnesia). Over his life, there was an increase in the number of years prior to the operation that he could no longer remember.

EUGENE PAULY

Eugene was born in 1922 and grew up in California, USA, eventually leaving home to become a radio operator for an oil company at sea. Upon his return he worked as an aerospace technician. By 1992 he was enjoying a peaceful retirement with his wife

THE BASAL GANGLIA

THE BRAIN AREA EUGENE AND HENRY USED TO LEARN NEW SKILLS IS FORMED OF THESE PARTS:

DORSAL STRIATUM
Made of the caudate nucleus and putamen, this is important for planning, carrying out and automating motor behaviour dependent on energy required and potential reward. This processing contributes to its role in forming habits and addiction.

VENTRAL STRIATUM
Containing the nucleus accumbens and olfactory tubercle, it evaluates risks/gains of behaviours.

GLOBUS PALLIDUS
This helps to regulate movement, motivation, reward, aversion and the control of actions and goals.

VENTRAL PALLIDUM
This receives strong signals from the brain's emotional circuits in order to govern reward and motivation behaviour and also produce motor outputs.

SUBSTANTIA NIGRA
This is important for regulating movement, and it receives information from the striatum directly and via the globus pallidus. It is the site of damage in Parkinson's disease.

SUBTHALAMIC NUCLEUS
This region of the brain is involved in making difficult decisions, plus some motor functions. It combines cognitive (thinking), emotional information and motor processes into an output.

when he suddenly contracted viral encephalitis – a virus spreading to his brain. The virus destroyed his medial temporal lobe and caused damage to areas in the wider temporal lobe by severing connections between brain cells. His resulting amnesia mirrored that of Henry Molaison, but the damage extending outwards through the temporal lobe also cost him his sense of smell, impaired his semantic knowledge and caused anterograde amnesia stretching back up to 50 years.

The word 'semantic' is defined as relating to meaning in language or logic, so semantic knowledge is an understanding of language and how it works, and semantic memory is remembering language, such as the names of objects. The logic aspect was tested by asking Eugene to identify and explain ambiguous sentences. Eugene performed worse on these tasks than 'control' participants (who had no brain damage), whereas Henry did not have any problems. This extra impairment for Eugene was due to the damage inflicted on his lateral temporal lobe, which Henry did not experience. Yet despite this slight difference in the areas damaged, the two displayed many similarities.

SIMILARITIES

Although the damage to their brains had been caused by very different circumstances, both men were left without medial temporal lobes. Therefore, neither Eugene nor Henry could form new memories, meaning they forgot events almost

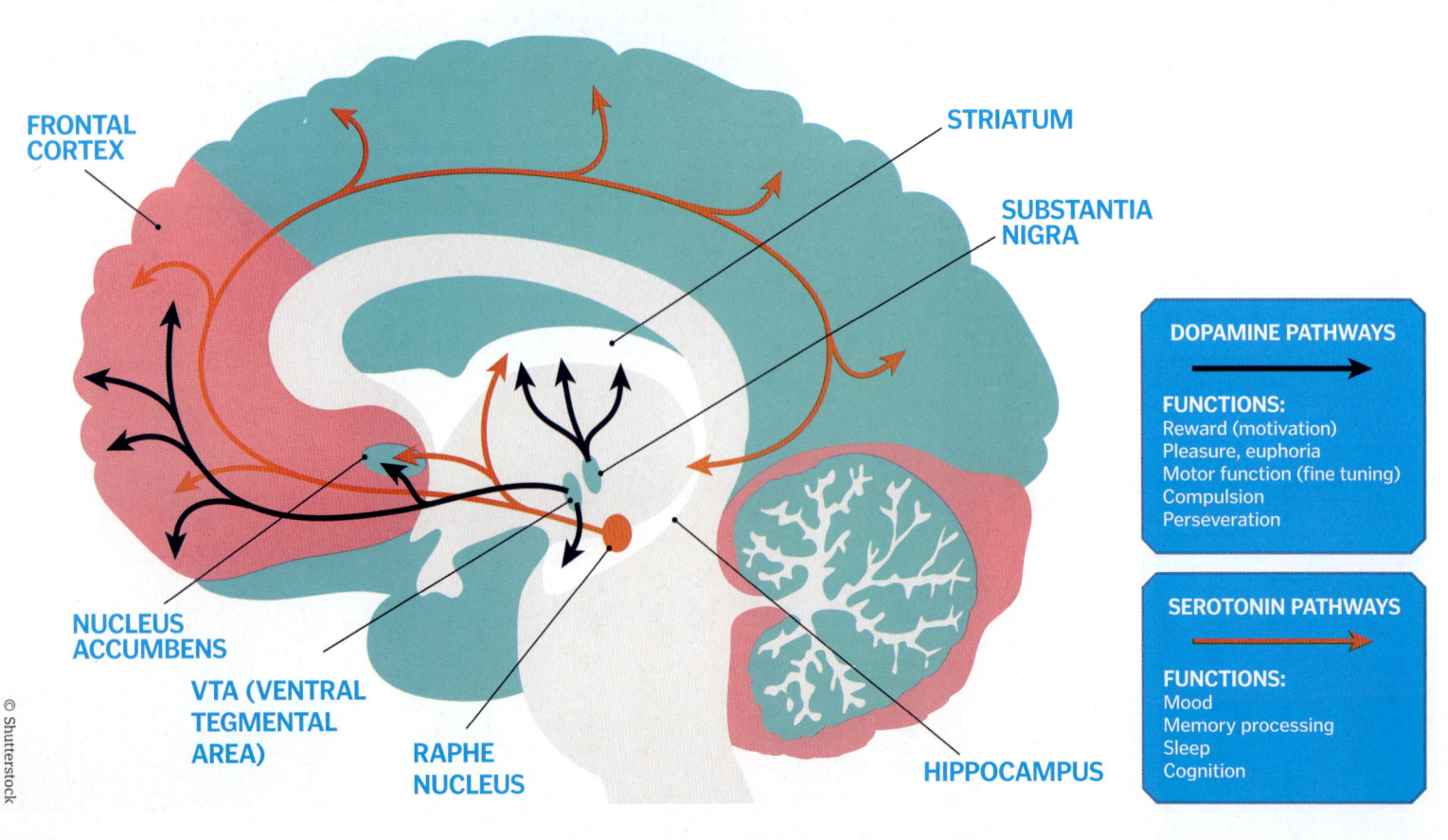

FRONTAL CORTEX

STRIATUM

SUBSTANTIA NIGRA

NUCLEUS ACCUMBENS

VTA (VENTRAL TEGMENTAL AREA)

RAPHE NUCLEUS

HIPPOCAMPUS

DOPAMINE PATHWAYS

FUNCTIONS:
Reward (motivation)
Pleasure, euphoria
Motor function (fine tuning)
Compulsion
Perseveration

SEROTONIN PATHWAYS

FUNCTIONS:
Mood
Memory processing
Sleep
Cognition

THE CURIOUS CASES OF HENRY AND EUGENE

instantly. Forgotten events included every new occurrence, from meeting someone new to eating breakfast. In fact, Henry encountered health problems because he would regularly forget that he'd eaten his bacon roll in the morning and repeat the task, resulting in him eating up to four of them for breakfast a day. Eugene would also sometimes have multiple breakfasts as it was part of his learnt morning routine, which he would forget he had already done.

The men's amnesia meant they were both confined to the present tense, frozen in a snapshot in time and permanently identifying themselves as being younger than they actually were. They would forget people as soon as they had been introduced and repeat statements they had just uttered as if it were the first time they were saying them. However, both men had normal intellectual and cognitive function aside from their memory issues. Although both remained anonymous for most of their lives, they were very willing to participate in research. The research, which provided a rare insight into how memory works, was initiated with Henry by PhD student Brenda Milner. Following this, more patients with medial temporal lobe damage, like Eugene, were thoroughly analysed by neuroscientist Larry Squire. This intensive research revealed some very specific learning tasks that rely on areas other than the medial temporal lobe.

AN ALTERNATIVE FORM OF MEMORY

This potential for learning demonstrates humans have distinct types of memory – declarative and nondeclarative. Researchers have since further classified these into more descriptive subtypes, but Henry and Eugene were pivotal in revealing these two initial distinctions. Declarative memory is the conscious ability to recall facts and events and is processed by the medial temporal lobe. Nondeclarative memory is unconscious and describes things such as motor skills, conditioning and habits. Due to their medial temporal lobe damage, it is declarative memory that Henry and Eugene lacked, but thorough testing revealed their capacity for nondeclarative learning in spite of their injuries.

A PIECE OF HENRY AND EUGENE'S BRAINS WAS MISSING, DUE TO VERY DIFFERENT CIRCUMSTANCES.

"The men's amnesia meant they would forget people as soon as they had been introduced"

TRICKS AND TRAUMA

NORMAL BRAIN

HENRY'S BRAIN

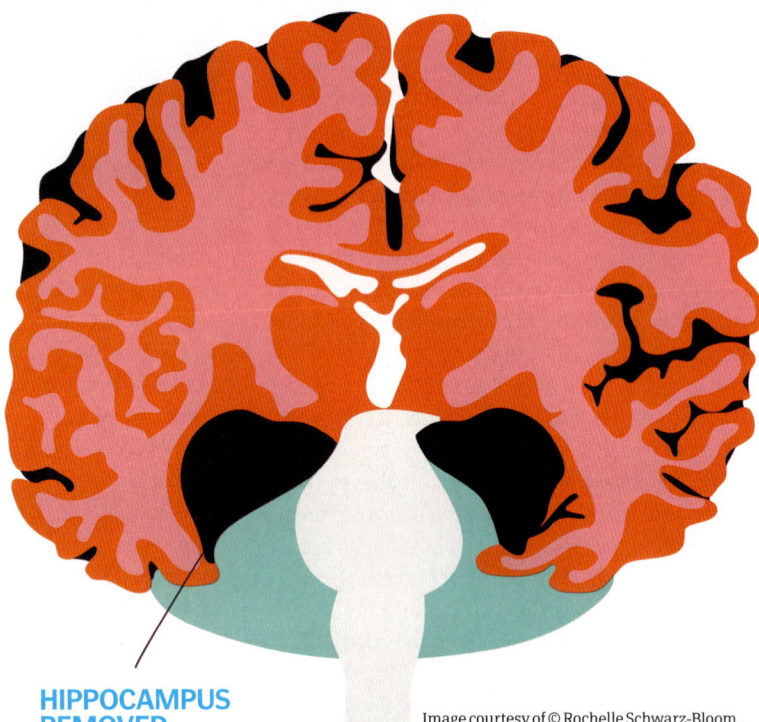

HIPPOCAMPUS

HIPPOCAMPUS REMOVED

Image courtesy of © Rochelle Schwarz-Bloom, Duke University. http://sites.duke.edu/apep

One such experiment was a memory task where two picture cards were presented to the participants (people with medial temporal lobe damage – also termed amnesiacs – and 'control' participants) in pairs. One card in the pair had 'correct' written on the back. The task was for the subjects to select the 'correct' card of the pair. The cards were always presented in the same pairs and the same card was always 'correct', so a pattern could be identified and memorised.

The amnesic patients were able to memorise this task, albeit more slowly than the controls. This suggested that they were learning the task using a part of the brain other than the medial temporal lobe, which had previously been attributed to all forms of memory.

The brain region responsible for this alternative form of memory was revealed by a group of patients with Parkinson's disease. Parkinson's disease is characterised by the death of neurons in the basal ganglia, an evolutionarily 'old' part of the brain that has roles including movement and reward behaviour. Parkinson's patients have intact medial temporal lobes, so their comparison with amnesic patients presents a double dissociation.

A double dissociation is the most conclusive way to show a particular brain area is responsible for a particular task. It compares a group of patients with damage in one area and an otherwise healthy brain with another group with damage to a different area but healthy rest of the brain. In this case, amnesic patients had a healthy basal ganglia but damaged medial temporal lobe, while

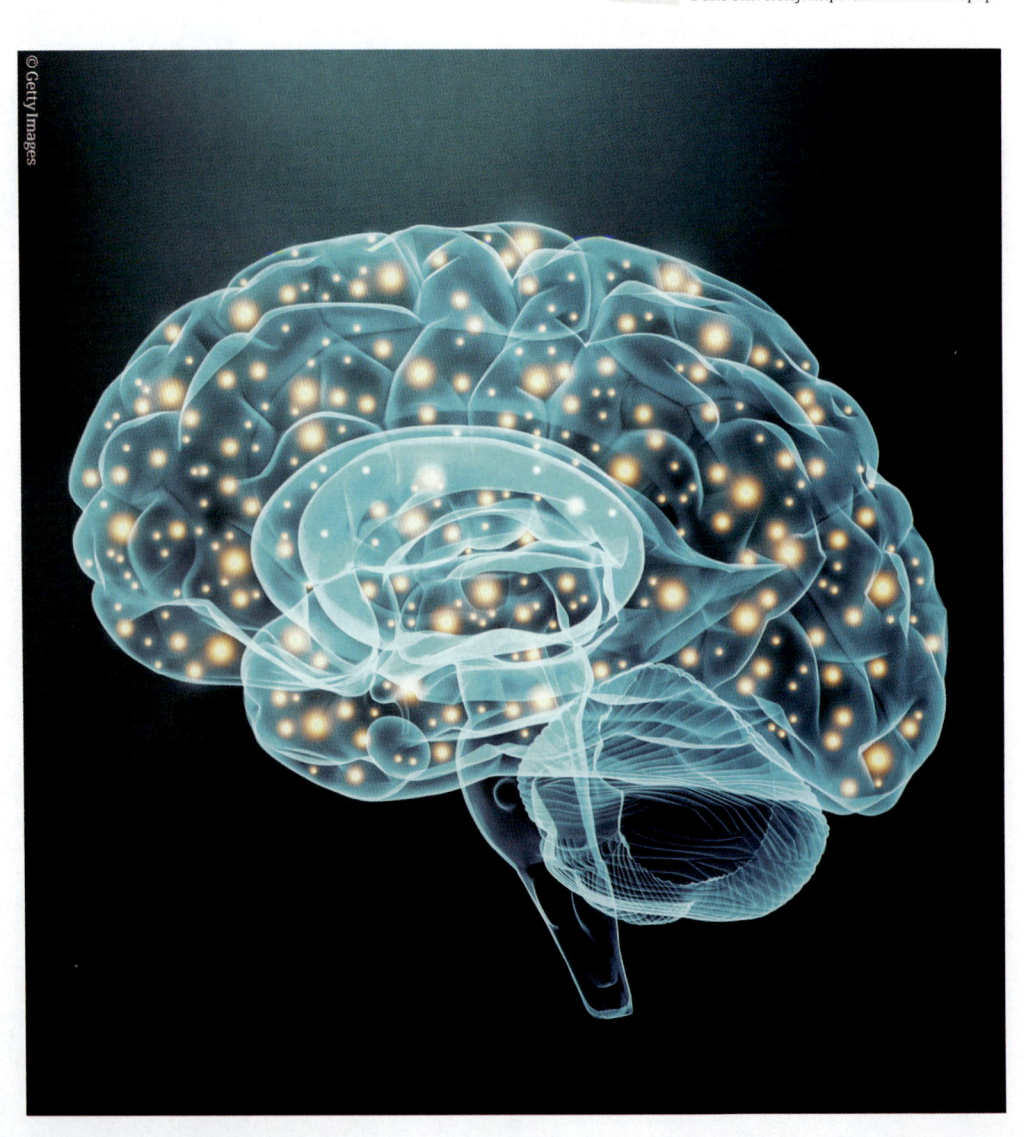

THE CURIOUS CASES OF HENRY AND EUGENE

Parkinson's patients had a healthy medial temporal lobe but damaged basal ganglia. The Parkinson's patients were able to perform tasks that tested their declarative memory but not tasks like the card-pairing task, which tested their nondeclarative memory. Conversely, amnesic patients performed well in the nondeclarative tasks but not the declarative tasks. Therefore, scientists concluded that the medial temporal lobe is in charge of declarative memory, while the basal ganglia is responsible for nondeclarative memory.

HABITUAL LEARNING

The nondeclarative memory of the basal ganglia is underpinned by a gradual process called habitual learning. This has now been shown in rats to occur in the neostriatum within the basal ganglia and is guided by the reward system. The reward system relies on chemical signals from the neurotransmitter dopamine. A habit is formed when an individual receives a reward for a certain behaviour that triggers a release of dopamine in

THE MEDIAL TEMPORAL LOBE

EUGENE AND HENRY'S MEDIAL TEMPORAL LOBES WERE BOTH DESTROYED. THIS PART OF THE BRAIN IS RESPONSIBLE FOR LEARNING AND TEMPORARY STORAGE OF MEMORY AND IS COMPRISED OF THE FOLLOWING:

HIPPOCAMPUS
Derived from the Greek for 'seahorse', the hippocampus is responsible for learning and making new spatial (mental maps) and declarative (facts and events) memories. It also links to the emotional centres contributing past experiences to current emotional events.

PARAHIPPOCAMPAL CORTEX
This area is responsible for episodic memory (personal experience) and visuospatial processing (navigating through a scene). These combine into contextual associations, i.e. using context and previous experiences to make sense of surroundings.

AMYGDALA
This combines information to produce emotional responses. It uses emotion to deduce the attractiveness or aversiveness of something.

ENTORHINAL CORTEX
This section consolidates memory with regards to sensory information by two-way communication with the hippocampus.

PERIRHINAL CORTEX
The function of this region is sensory perception that enables an individual to recognise and memorise objects. It transmits sensory information to and from the hippocampus via the entorhinal cortex.

SUBICULAR COMPLEX
With its name taken from the Latin for 'support', the subicular complex receives and processes input from other brain areas, e.g. it amplifies signals from the hippocampus. The front of the subiculum mainly processes memory, movement and spatial information, while the back can soften the body's stress response.

© Getty Images

TRICKS AND TRAUMA

the basal ganglia, motivating them to do it again. In animal models, this reward is usually a piece of food, but it can be anything, such as the human participants seeing the word 'correct' and being congratulated by the experimenter when they turn over the right picture card.

These subtle rewards are how the amnesic patients Henry and Eugene developed habits in their daily lives following their brain injuries. The rewards become the goal and create the motivation to perform the goal-driven behaviour over and over until it becomes a learnt, subconscious habit. For example, Eugene was able to go to his kitchen to get food when he was hungry, as the food was the reward, but he was not able to recount where the kitchen was when asked, as this yielded no reward. Similarly, he learnt a specific route near his home where he would go for a daily walk and find his way home again, despite not being able to consciously identify which house was his.

The key to habitual learning is the cues remaining identical. If a tree or roadworks blocked his path, Eugene would be lost and unable to navigate home because the habit had been disrupted. Similarly, if the picture cards were presented in different pairs or all together, the amnesic patients were unable to identify the 'correct' cards.

Another example are the behavioural changes Eugene showed over the 14 years he was studied. When experimenters first visited him he was hostile and reluctant to be tested, requiring persuasion from his wife. However, after several years of the same researchers visiting him, he greeted them warmly and instinctively went to the table he was usually tested at, even without his wife's presence. Interestingly, he still denied ever having met them or taking part in these tests before.

This change in behaviour illustrates habitual learning by Eugene subconsciously and gradually over several years as a result of regular occurrences in a specific environmental context. Therefore, habitual learning is a very specific form of learning that is intrinsically dependent on ritual and repetition.

Repetition also comprised another form of memory Henry and Eugene were able to exhibit: working – or immediate – memory. Through repeating a string of numbers, they could retain the sequence in their minds for up to 20 seconds.

However, if their attention was drawn elsewhere they would instantly forget they had even been tasked with remembering numbers. This demonstrated that this working memory was being stored temporarily in an area other than the medial temporal lobe. Research into attention and working memory has pointed to the involvement of several regions, from the prefrontal cortex to areas deep within the brain.

All of these different forms of memory that Henry and Eugene retained show just how complex and nuanced process memory is. They provided an invaluable insight into the brain's ability to find novel ways to function when the ordinary route is blocked.

This incredible ability was cemented by the fact that Henry managed to live a relatively functional life for 55 years without a medial temporal lobe. Both men's incredible brains are now stored at the Brain Observatory in San Diego, USA, and continue to contribute to neuroscientific knowledge to this very day.

They may not be known all that well in the wider world, but within the field of neuroscience, Henry and Eugene's legacy will live on forever.

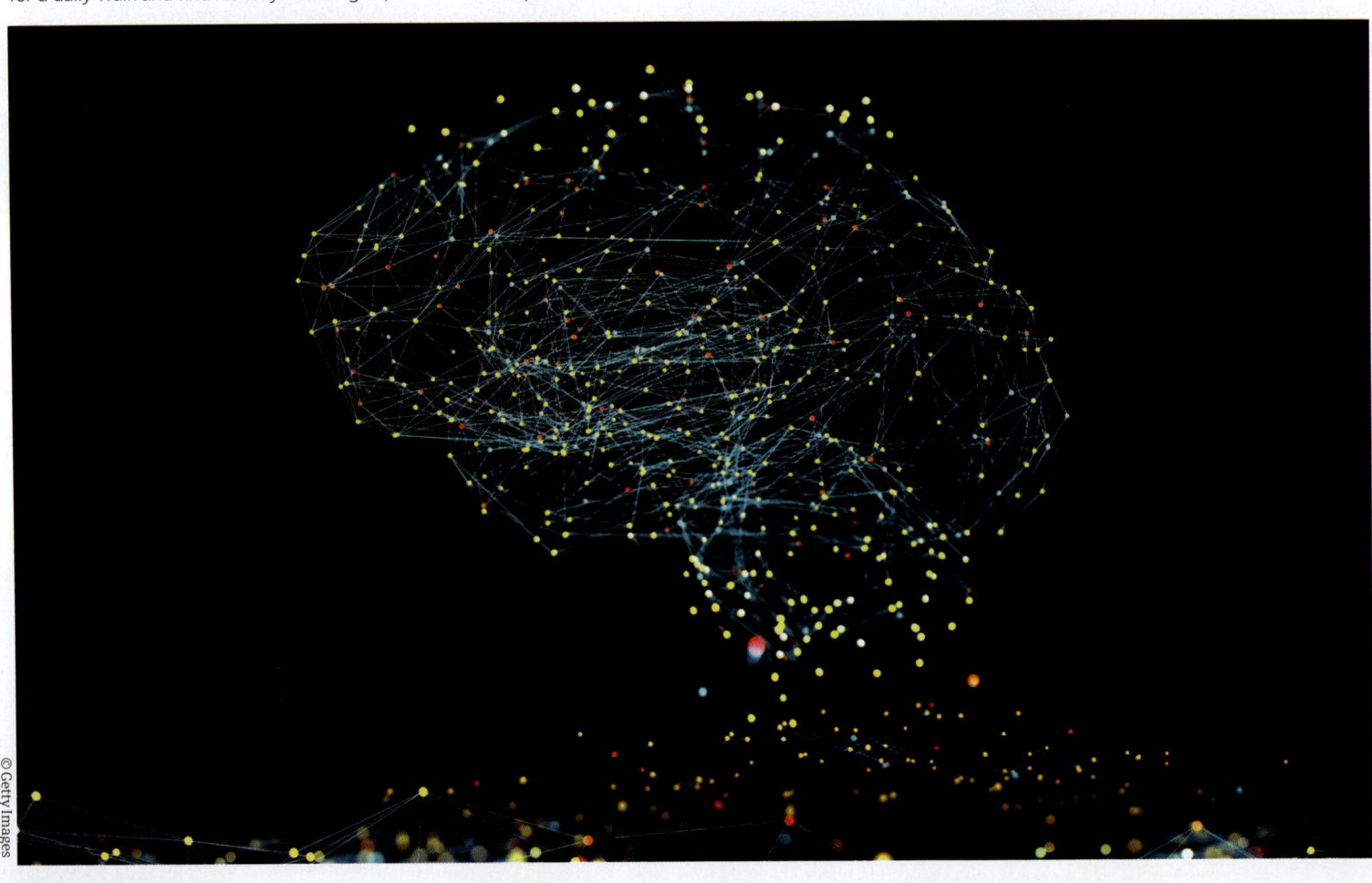

> "Eugene was able to go to his kitchen to get food, as the food was the reward, but was not able to recount where the kitchen was when asked"

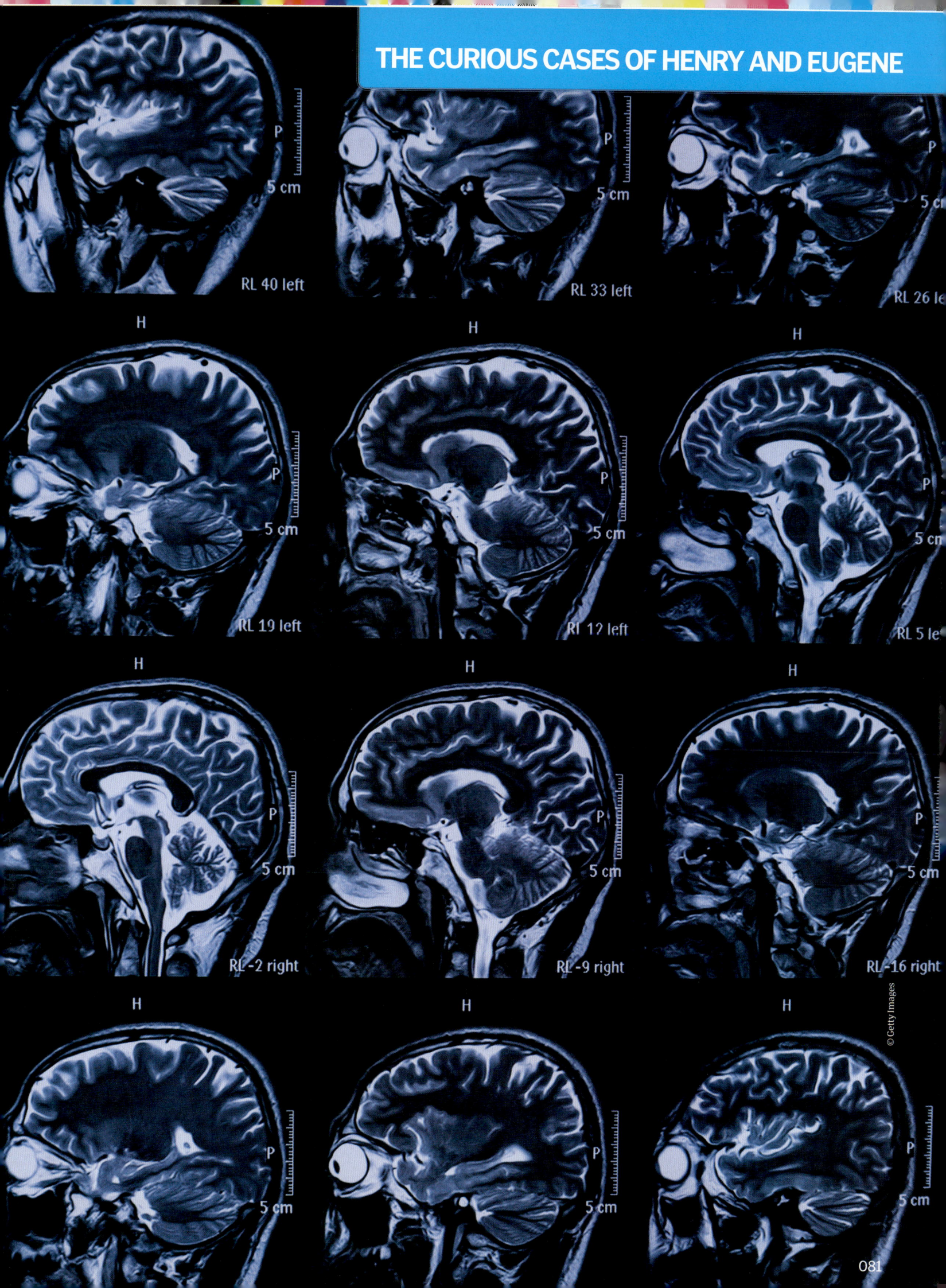

THE CURIOUS CASES OF HENRY AND EUGENE

TRICKS AND TRAUMA

THE CRUELLEST DISEASE

THE VARIOUS FORMS OF DEMENTIA NOW AFFECT 50 MILLION PEOPLE AROUND THE WORLD. THERE ARE 10 MILLION NEW SUFFERERS EVERY YEAR – AND STILL NO CURE

Words by **Edoardo Albert**

Although there is not much good to be found in dementia, its increasing prevalence is a consequence of something that is undeniably good: we are all living longer. For while there are a number of risk factors associated with dementia, by far the most important is age. Since people are living longer, it stands to reason that more of us run the risk of developing dementia.

In 1850, the average life expectancy for a European was 36.3 years, while in 2014 it was 86.7 years. There is only a one-in-a-thousand risk of anyone under 60 developing dementia. Before the astonishing increase in life expectancy that we have seen over the last 150 years, there would have been few cases of dementia as few people lived long enough to develop it. But now, with an increasing proportion of the population living into their 90s and beyond, the number of people at risk from dementia has increased. For people between 60 and 64, one in a 100 will probably develop dementia. For those between 75 and 79, that increases to six in 100. Between 90 and 94, the risk increases to 30 per cent, and for those who live to 95 and older the risk is 41 per cent. Dementia is fundamentally a disease of aging.

WHAT IS DEMENTIA?

There are a range of conditions that fall under the general label of dementia. The most common is Alzheimer's, which affects between 50 and 70 per cent of dementia sufferers. The various dementias impact the brain's ability to think and recall, having a dramatic effect on a person's ability to perform everyday tasks. In its most advanced forms, dementia leaves the sufferer unable to survive without constant care.

After Alzheimer's, the next most common form of dementia is vascular dementia. This condition can arise after suffering a stroke and accounts for about a quarter of dementia cases. Other relatively common forms include dementia with Lewy bodies and frontotemporal dementia.

A rarer type of dementia can be caused by normal-pressure hydrocephalus, which results from excess cerebrospinal fluid. This condition presents a triad of symptoms causing gait abnormalities, incontinence and dementia (often called, in a diagnostic mnemonic, 'wet, wacky and wobbly'). Several other disorders (including Parkinson's and Creutzfeldt-Jakob disease) and even some infections can also cause dementia.

All of these conditions cause the deterioration in brain function and memory that are collectively known as dementia. One of the difficulties that has bedevilled the search for treatments is the

> "A range of conditions fall under the general label of dementia"

lack of a satisfactory diagnostic test for the various different forms of dementia – particularly for Alzheimer's. Diagnosis is based on a medical examination, tests and a case history and requires significant impairment in at least two of the five key mental faculties: memory; language and communication; concentration and attention; judgement and reasoning; and visual perception. However, since the symptoms of the different forms of dementia overlap considerably, it is often difficult for doctors to diagnose the specific form of dementia from which the patient is suffering.

RISK FACTORS FOR DEMENTIA

Above everything else, the chief risk factor for developing dementia is age. This is particularly true for the most common form of dementia, Alzheimer's. However, after the complex effects of ageing, other factors come into play. One of these is having a family history of the disease, which increases the risk of inheriting genes that predispose people to developing Alzheimer's.

There is also a strong association with Down's syndrome. Both Down's syndrome and Alzheimer's are associated with abnormalities in chromosome 21. People suffering from Down's syndrome have a higher-than-average risk of developing Alzheimer's as they age, with close to half of Down's syndrome sufferers developing Alzheimer's in their 60s (whereas it's less than ten per cent in the wider population).

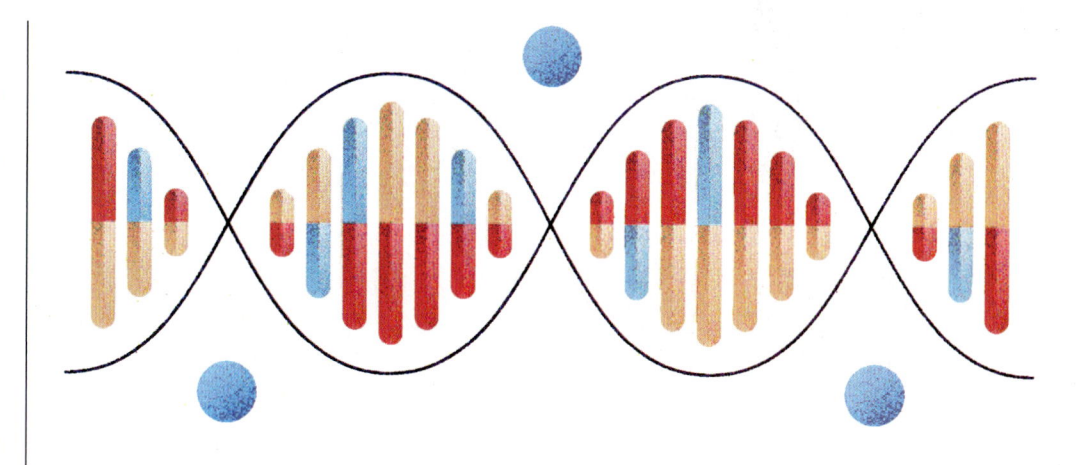

Apart from Down's syndrome, other cognitive impairments also increase the risk of dementia. While these factors are beyond personal control, others can be guarded against. For instance, having a history of head injuries significantly increases the risk of dementia. Medical advice regarding the effects of concussion is changing – the old days of playing rugby and being sent back on the field as long as you could stand up are thankfully past. But low-intensity brain injuries can have a cumulative effect, resulting in dementia. Sports that could potentially involve head injuries should be performed with helmets where possible and, crucially, sufficient time must be given for people to recover from any concussion.

The classic signs of unhealthy living – smoking, obesity and drinking to name a few – are also implicated in the development of dementia. Unsurprisingly, the medical advice is to stop smoking, keep fit and drink in moderation. Not only will this reduce the risk of dementia, but it will significantly improve your overall health as well.

On the health front, it's not just the unholy trinity of smoking, drinking and obesity that can lead to dementia – our mental health can play a role too. Studies have indicated that people over 55 who suffer from steadily increasing levels of depression are at greater risk of developing dementia than those with low or variable levels. Wider studies have also shown that social isolation, which often increases as people get older, is linked to a greater vulnerability to dementia. In fact, the lonelier and more isolated someone becomes as they get older, the greater their risk of developing dementia.

IS THERE ANYTHING WE CAN DO TO AVOID GETTING DEMENTIA?

The good news is that the things we can do to reduce our risk of developing dementia are generally beneficial to our overall health and well-being. Ever wanted to learn a new language or how to play the guitar? Do it. It doesn't matter how old you are, activities that require intense concentration (preferably paired with new muscle memories) are ideal. Learn to tango. Become a potter. Make ships out of matchsticks. Sing, or better yet, join a choir. That way the learning will be combined with fresh and exciting social relationships. Human beings are social animals and, as a general rule, we need company to function properly. Solitary confinement is used as a punishment – don't let it become a lifestyle. Aside from these activities, the medical conditions that can lead to strokes – and hence vascular dementia – should also be guarded against. High among these risk factors is diabetes, so exercise and a healthy diet (which help reduce the chances of developing type 2 diabetes) can therefore reduce the risk of getting dementia.

THE GENETICS OF ALZHEIMER'S

Within the many variations of dementia, most research has been focused on investigating the genetics of Alzheimer's. The strongest evidence for a genetic cause has been found for a variant called early onset Alzheimer's disease. As the name suggests, this devastating form of the main syndrome manifests far earlier than the more common late-onset form, affecting people in their 30s to mid-60s. This variant affects less than ten per cent of all those who get Alzheimer's.

Of those who suffer from early onset Alzheimer's there is a small subset, affecting about 500 families around the world, where the disease is caused by mutations in chromosomes 1, 14 and 21. So within a very small section of the population of people with Alzheimer's – probably less than one per cent – a direct genetic cause for the disease has been found.

Scientists have not yet discovered a direct trigger gene for the more common late-onset form of Alzheimer's. However, a gene in chromosome 19 called apolipoprotein E (APOE) has been found to play a role in the disease.

Of the different alleles (variants) of the gene, the most common – APOE 3 – is neutral with respect to Alzheimer's. On the other hand, the allele APOE 2 may protect against Alzheimer's, while APOE 4 is a risk factor for developing the disease. However, with Alzheimer's nothing is ever straightforward.

TRICKS AND TRAUMA

While carrying the APOE 4 allele in chromosome 19 increases the risk of developing Alzheimer's, and at a younger age, it is not a necessary and sufficient condition. Some people with the risk allele never develop the disease, and others who do get Alzheimer's do not have the APOE 4 allele.

Researchers have widened the search for genes that might cause Alzheimer's using genome-wide association studies, which look for links between specific genes (or gene variants) and certain traits. These projects have revealed more regions in the genome (the complete DNA of an organism) that appear to be associated with the disease. Further research is needed to find out what roles these regions play in the development of Alzheimer's.

WHAT DOES ALZHEIMER'S DO TO THE BRAIN?

Nothing good. The disease was first identified by Alois Alzheimer, a German psychiatrist, in 1901. A woman named Auguste Deter, a patient at an asylum in Frankfurt, came to the attention of Alzheimer. During the next five years, until she died in 1906, he became fascinated by her case. Asked questions, Deter would reply, "Ich habe mich verloren" ("I have lost myself").

When Deter died, Alzheimer examined her brain. He found it contained clumps of protein called plaques. We now know that these plaques are made up of beta-amyloid protein and are a key feature of the disease. As Alzheimer's progresses it destroys brain cells (called neurons) and disrupts the neurons' connections with each other, a vital aspect of mental function. Usually, the disease first attacks the neurons in areas of the brain involved with memory before spreading to the cerebral cortex and those areas responsible for reasoning, language and behaviour. In someone with late-stage Alzheimer's, the brain will be atrophied, with the specific regions affected by Alzheimer's reducing in volume.

Such major physical changes in the brains of patients with Alzheimer's are associated with many microscopic and biochemical changes in their brains too. The beta-amyloid plaques that Alzheimer first identified form clumps in between neurons, disrupting their function. Research also suggests that the beta-amyloid plaques interact with another feature of Alzheimer's: tangles of tau proteins. These tangles form within the neurons and prevent nutrients being transported to the extremities of the cells properly, and they stop neurons communicating properly. It seems that abnormal tau proteins gather in memory-related parts of the brain, while beta-amyloid plaques form between the neurons in these regions. When the amount of beta-amyloid plaques reaches a critical level, a cascade of abnormal tau proteins spreads through the rest of the brain, causing neuronal collapse.

Other research has pointed to chronic inflammation in the brain being another cause of Alzheimer's. The cells that normally clean away accumulated junk chemicals in the brain (the microglia and astrocytes) fail to do their job and – as a double whammy – start to produce a chemical that causes chronic inflammation and further neuronal damage.

WHAT ARE THE PROSPECTS FOR EFFECTIVE TREATMENTS OR EVEN A CURE?

In the short term, poor. We still lack an adequate understanding of what causes most forms of dementia. The causes of Alzheimer's are proving particularly opaque. It was hoped that drugs designed to remove the build-up of beta-amyloid plaques would treat or even cure the disease, but these initial hopes have been dashed. Multiple drug tests aimed at treating the beta-amyloid plaques have failed, but a number of alternative theories regarding the cause of Alzheimer's have been advanced recently, including environmental toxins, inflammation and bacterial, viral and fungal infections.

A possible implication of this multiplication of proposed causes is that Alzheimer's might not be a single disease, or that it is a single disease but with many different causes. If Alzheimer's is actually a number of different conditions that exhibit similar symptoms, then researchers will have to find a way to distinguish the different forms of the disease before they are able to develop specific treatments for each type. It is a complex and time-consuming task, but if Alzheimer's should prove to be a single disease with multiple causes – even if those causes are

---

FROM THE FRONT LINE OF DEMENTIA RESEARCH

We spoke to Dr Giovanna Lalli, the UK Dementia Research Institute director of scientific affairs, to ask her about the future direction of dementia research. "It's very important not to think in silos; at the UK Dementia Research Institute we're investigating different types of dementia, which are characterised by inflammatory states, by misfolding or the aggregation of toxic proteins. We're tackling dementia from different angles and perspectives – from genetics to the interactions between brain cells, vasculature and the immune system, to the identification of novel biomarkers. We want to be innovative and drive a step change in how we understand these diseases in order to accelerate the discovery and delivery of interventions to diagnose but also treat and ultimately prevent dementia. This is our ultimate goal – that's why we exist. We're gaining momentum, and the fact we're able to attract young talent from across the world is exciting – we need fresh thinking, fresh perspectives to tackle this problem.

"I'm optimistic we're going to have breakthroughs in our knowledge over the coming years, because we have new methods of studying how different types of cells in the brain interact and how different genes are signalling to each other. By using the latest technologies – such as high-resolution microscopy and single-cell sequencing, cutting-edge neuroimaging and probes able to record activity in hundreds of different neurons simultaneously – and harnessing the power of all these techniques together with insights coming from genetic studies can help us better understand not only how the brain works, but also how dementia develops. Then it will be a case of translating these findings into treatments."

DEMENTIA

harder to identify – a treatment might then be easier to find.

THE IMPORTANCE OF CARE

With a cure for Alzheimer's and other forms of dementia still a long way off, the importance of how we care for those living with these conditions grows more apparent. In many cases, the carer is the spouse of the patient, a final service paid in honour and love. But it is not easy.

We spoke to Dementia UK's Pat Brown, who works on their Admiral Nurse Dementia Helpline, to learn more about the importance of caregiving and the needs of carers.

"With the population living longer, there are more diagnoses of dementia – it is now one of the biggest health issues of our time. As such, there urgently needs to be more support, not just for the person diagnosed, but also for the often-silent generation of caregivers. They experience first-hand the devastating changes this condition can bring for a partner or family member.

"Dementia UK's Admiral Nurse Dementia Helpline takes many calls from people facing dementia, including from carers in distress. When talking to carers, Admiral Nurses can provide unique expertise and experience to a condition that has over 200 subtypes. They can get to the heart of carer's challenges, allowing them to see the importance of making time for themselves and accessing the right support for their circumstances."

If you are a carer, or know a carer who is struggling, or require any help and support with dementia, ring the Dementia UK helpline on 0800 888 6678. The helpline is staffed exclusively by experienced Admiral nurses.

A POSSIBLE TREATMENT FOR ALZHEIMER'S?

Amid the general gloom over treatments for Alzheimer's, one new approach has improved the symptoms of some patients. Dr Dale Bredesen – working on the belief that Alzheimer's is not a single disease but has at least three different varieties – claims to have identified 36 different factors that contribute to the various forms of Alzheimer's. For a treatment to work, Dr Bredesen and his associates argue that the patient has to be assessed thoroughly to see which combination of these factors are implicated in causing their Alzheimer's, then the causative factors have to be removed or ameliorated. According to Dr Bredesen, there are three groups of causative factors: inflammation, which may be caused by a combination of infections, diet and other causes; toxins, including moulds and chemicals; and a decrease in the level of hormones and nutrients in the body.

Once the combination of factors affecting the patient have been identified, a detailed programme of exercise, diet, hormones and supplements is prescribed. While the programme has produced an improvement in some of the patients, it requires ongoing commitment for the beneficial effects to continue. Those who have stopped the programme have reported a return of Alzheimer's symptoms, so it can't be considered a cure for the disease. However, it is one of the very few treatments that has produced any improvement in symptoms. Further research and proper clinical trials are required to evaluate the treatment's full potential for future use in the battle against this most devastating of diseases.

SUSIE AND THE SAVANTS

96 SUSIE MCKINNON

88 SUPER SAVANTS

SUSIE AND THE SAVANTS

100 PHOTOGRAPHIC MEMORY

104 PAST LIVES

102 GENERATIONAL MEMORY

SUSIE AND THE SAVANTS

STEPHEN WILTSHIRE
LANDSCAPE MEMORISER

Londoners see their city skyline every single day, but if they had to draw a picture of it, most would find themselves struggling to recall even half of the details. Autistic savant Stephen Wiltshire can create an accurate image of every building entirely from memory.

He didn't say a word until the age of five and only started talking properly at the age of nine, but he always had a talent for art. He sold his first piece – a painting of Salisbury Cathedral – to the British prime minister at the age of eight, and by the time he was 13 he'd published a book of sketches titled *Drawings*.

Wiltshire only needs to see a cityscape for a few minutes to be able to reproduce it from memory. He works on enormous canvases, filling in the black outlines of hundreds of buildings, streets and landmarks entirely from his mind's eye. His incredible talent has seen him tour the world to tackle famous skylines, including London, Paris, Edinburgh, Venice, Amsterdam, Moscow, Tokyo, Chicago and New York.

SUPER SAVANTS

THE REAL-LIFE
RAIN MAN
AND OTHER SUPER SAVANTS
A LOOK INSIDE THE BEAUTIFUL MINDS OF SOME OF HISTORY'S MOST EXTRAORDINARY INDIVIDUALS

Words by **Laura Mears**

Savants have some of the most phenomenal minds in human history. Changes to the wiring of their nerve cells enable them to use their brains in ways that others can only dream of. These rare individuals have autism, developmental disorders, or brain injuries accompanied by some truly incredible and unusual talents.

Savants are able to access raw information that ordinary minds don't normally get to use. They have deep and powerful memories, and they seem to be able to find hidden rules unconsciously, allowing them to perform incredible feats. These tend to focus on five main skill areas: music, art, mathematics, spatial reasoning and calendar calculations. Each one is underpinned by rules that require immense effort for most of us to master. But savant minds are able to absorb musical scales, artistic geometry and mathematical logic with ease.

Within each specialist area there are three levels of savant ability. 'Splinter skill savants' have encyclopaedic knowledge of a highly specialised interest. They're often intensely focused on one subject area, perhaps dates, timetables or phone numbers. 'Talented savants' combine knowledge with a remarkable skill, expressing abilities in music, maths or art that excel far beyond those of other people. 'Prodigious savants' are the rarest of all, with talents so outstanding that they fall into the category of 'genius'.

The ten remarkable individuals featured here are some of the most famous savants in human history, with talents ranging from a photographic landscape memory to a minute-perfect internal metronome.

SUSIE AND THE SAVANTS

WHAT IS SAVANT SYNDROME?
TRYING TO UNDERSTAND THE REASON BEHIND A SAVANT'S SKILLS

Savants have deep memories and astonishing cognitive abilities, but the science underpinning their skills is still something of a mystery. There are two main schools of thought that attempt to explain what's going on.

The first describes savant syndrome as a form of hypermnesia – a brain with an unusually vivid memory. The second explains it using a theory called 'weak central coherence' – a brain that struggles to put information into context.

Imagine a group of blindfolded people touching different parts of an elephant. The one at the trunk might imagine they feel a snake. The one at the ear, a fan. The one at the leg, a tree trunk. And the one at the tail, a rope. It's only by taking the blindfolds off and looking at all the parts together that reality becomes clear.

Savants often lack the ability to organise and generalise information. Rather than searching for meaning in data, their brains focus on memorisation and direct recall. They see the parts of the elephant individually, rather than taking a step back to look at the whole.

Ordinary brains see the whole, savant brains see the parts

KIM PEEK
REAL-LIFE RAIN MAN

In the 1988 film *Rain Man*, Dustin Hoffman played autistic mega-savant Raymond Babbitt, a man with an extraordinary memory. He could instantly recognise a waitress's name and recite her phone number by heart, having memorised the telephone directory the night before. Kim Peek was the savant who inspired the film, and this real-life Rain Man could do the same.

Peek was born with a problem in a part of the brain called the cerebellum, which controls voluntary movement. He also lacked a nerve bundle called the corpus callosum, which connects the left side of the brain to the right. These differences made it hard for Peek to coordinate his movements but gave him unprecedented access to his long-term memory.

Peek was able to read both pages of an open book at once, one with his left eye and one with his right, taking only ten seconds to commit each spread to memory. By the time of his death in 2009, he had memorised the contents of over 12,000 books. His vast knowledge repertoire included history, literature, music, sports, dates and zip codes.

SUPER SAVANTS

ORLANDO SERRELL
HUMAN CALENDAR

At the age of ten, a sporting accident changed Orlando Serrell's life forever. He was running for first base when a baseball hit him on the left side of his head, causing a headache that lasted for days. He decided that he wouldn't tell his parents about the headaches, and no one noticed anything was different. That was until the pain started to subside. At that moment, Serrell found he had became a human calendar.

Other people with human calendar abilities have memorised the 'rules' of days and dates, allowing them to work out the day of the week for any date of any year. But for Serrell it's more than that. He has hyperthymesia, which literally means 'excessive remembering'. He says that he can 'see' a calendar of his life in his mind, which enables him to recall exactly where he was, what he was doing and what the weather was like on any day since he had his accident.

JEDEDIAH BUXTON
MENTAL CALCULATOR

Jedediah Buxton was one of history's first recorded autistic savants. Born in Elmton in Derbyshire, UK, in the early 1700s, his father was master of the local school, but Buxton never attended. He didn't learn to write, he couldn't read and he didn't understand music, but he was without doubt a mathematical genius.

As a boy he spent his time outdoors, pacing local fields to work out their size. He could accurately measure distance using nothing but his stride length and then transform the measurements into acres, inches or hair widths using only his mind.

For most of his life Buxton worked as a farm labourer in Derbyshire, but at the age of 47 he decided to walk to London. Once in the capital, he showed off his mathematical skills to an audience of stunned scholars at the Royal Society, the oldest scientific academy in the world.

Afterwards, he visited the Drury Lane theatre to see a Shakespeare play, but he couldn't enjoy the show. He ended up completely absorbed in counting the lines spoken by the actors and the steps taken by the dancers.

SUSIE AND THE SAVANTS

TEMPLE GRANDIN
ANIMAL WHISPERER

Dr. Temple Grandin is a professor of animal science, an author and a world-leading expert on autism, but she didn't speak until she was three years old. Rather than think in words, she thinks in pictures. She says that this allows her to understand animals because they think in pictures too.

Scans of Grandin's brain have revealed that the bundles of brain cells running to her visual cortex are twice as thick as the bundles in the brains of other people. She calls them her 'internet trunk line for graphics'.

She argues that language covers up the visual thinking parts of the mind. Without words, she's free to think with her visual cortex. Using these skills, she can be hypersensitive to changes in her environment, noticing things that no one else sees. She spotted that cows were afraid to walk over shadows and that they react differently to people on horseback versus people on the ground. These visual skills have allowed her to design farm facilities that reduce animal stress, and they're now used all over the world.

SUPER SAVANTS

LESLIE LEMKE
MUSIC MAESTRO

Leslie Lemke only has to hear a piece of music once to memorise it. It's a skill he's had since he was a boy. He was born prematurely and problems with his eyes left him blind at just a few months old, so he has never seen music written down. But, by using his highly attuned hearing, he's able to pick out a melody from a complex soundtrack and play it back on the piano from memory.

The muscles in Lemke's hands are stiff, so he has trouble with the fine movements needed to do everyday tasks, but when he sits at a keyboard, his fingers relax. And once he's in the zone his musical talents are astonishing. Not only can he repeat a song back after only one listen, but he can play along with a tune he's never heard. He can hear each note, process it and repeat it with barely a second's pause. As with Alonzo Clemons, the media storm stirred up by *Rain Man* propelled Leslie Lemke into the limelight, exposing his talent to the public.

ALONZO CLEMONS
SUPER SCULPTOR

Alonzo Clemons acquired a remarkable gift for modelling after sustaining a head injury at the age of three. At just five years old he began to sculpt model animals from household materials like soap and lard. By the time he was ten he was working with clay, using his hands and fingernails to create realistic figures at high speed.

His head injury affected his development, and he spent all of his teenage years in an institution. His teachers tried to encourage him to learn other skills by taking away his clay, but he wasn't interested in anything but modelling. He'd search the building for anything to model with, from moulten tar to window putty.

Clemons spent 20 years honing his craft before the release of the film *Rain Man*, and he credits the subsequent media interest in savants with his rise to fame. Now he works as a professional sculptor, and limited-edition bronze casts of his models sell all over the world.

ARE WE SAVANTS IN DISGUISE?
OUR BRAINS COULD BE SUPPRESSING THE SAVANT ABILITIES WITHIN US

Savants have incredible memories. Not only can they store more factual information than the rest of us, but they can also access it in its raw, unprocessed state. Some are born with their skills, others acquire them after a brain injury, and a select few become savants overnight for no apparent reason at all. These 'sudden savants' suggest that savant syndrome might lie dormant in us all.

World-leading expert in savant syndrome Darold Treffert suspects that ordinary brains suppress their natural savant tendencies. If you try to recall a distant memory, like every member of your secondary school class, you might not be able to retrieve the information, but that doesn't mean it isn't there. Look at a picture from your yearbook and forgotten details will soon come flooding back.

It's not that ordinary brains don't make memories like savant brains, it's just that we can't access them as easily. Something about savant syndrome unlocks the mind's memory capabilities, allowing access to storage that's normally hidden.

Memories of your childhood are still there, they're just locked away

SUSIE AND THE SAVANTS

Interfering with the left side of the brain triggers savant-like skills in ordinary minds

THOMAS WIGGINS
MELODY MIMIC

Thomas 'Blind Tom' Wiggins was born into slavery in 1800s America. Almost killed by his master for his disabilities, he was auctioned off as an infant and ended up on a farm belonging to General James Bethune.

As a young child, Wiggins wasn't able to speak for himself, but he became adept at mimicking sounds. He could repeat the crow of a cockerel, the bleat of a lamb and even the conversations that he heard around him. Eventually, Wiggins heard one of Bethune's daughters playing the piano, and it opened a door to a whole new world. By the time he was six he was performing his music in public.

Wiggins' savant skills were extraordinary – he had perfect pitch and could repeat any composition note for note – but he was exploited at every turn. At the age of eight, Bethune licensed him to a travelling showman. Then, as an adult, the family took him on tour in North America and Europe. By the end of his career, Wiggins had mastered more than 7,000 different songs, including several of his own compositions. But he never saw a penny of his income.

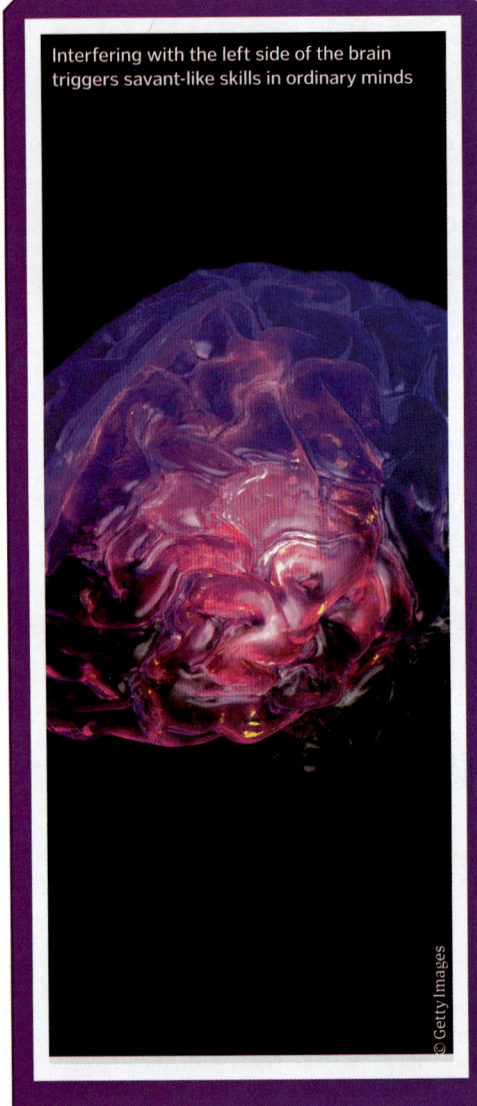

CAN YOU LEARN TO BECOME A SAVANT?

HOW MAGNETIC PULSES CAN TRICK THE BRAIN TO INDUCE SAVANT SKILLS

While it's not possible to learn to become a savant, that doesn't mean that we can't all experience savant-like skills. Savants tend to focus on the parts rather than the whole, ignoring concepts in favour of details, and studies suggest that this might have something to do with part of the brain called the left anterior temporal lobe.

The left anterior temporal lobe plays a crucial role in semantic memory – the memory of concepts. Using magnetic pulses to interrupt this part of the brain in non-savants seems to be able to induce savant-like skills. The left side of the brain normally searches for patterns, and without it people become more detail focused. Drawing style changes, proofreading ability improves and estimates of large numbers become more accurate. Blocking the left side of the brain seems to obscure the meanings we attach to things, preventing us from being distracted by the bigger picture and letting our hidden savant skills shine through.

BLIND TOM.
vright 1880, by John G. Bethune.

SUPER SAVANTS

ELLEN BOUDREAUX
HUMAN METRONOME

Ellen Boudreaux has never seen a clock, but she can tell you the time down to the minute. When she was born there was a problem with her eyesight, and over the first few months of her life she went blind. She developed acute hearing to compensate and, by the time she was six months old, her sister noticed her humming along with a lullaby toy.

Boudreaux later started repeating tunes on an electric organ, and when she was seven her teachers encouraged her to learn the piano. Then something amazing happened.

Boudreaux had been afraid to use the telephone, so her mum suggested that she practised by calling the talking clock. This automatic service provided the time of the day in hours, minutes and seconds. After listening to the call, Boudreaux developed an internal clock of her own.

Now, not only does she know exactly when her favourite TV shows are about to start, she can use her internal metronome to create astonishing music. She can break down orchestral pieces and recreate them as piano solos. What's more, she can make up accompaniments to music she's never heard before.

GOTTFRIED MIND
RAPHAEL OF CATS

Most famous for his images of cats, Gottfried Mind had a special talent for drawing fur. He spent most of his adult life hunched over a desk in a room filled with cats, kittens and rabbits, whiling away the hours creating realistic images of animals and children entirely from memory.

Born in 1768 in Switzerland, Mind was small and weak as a child. He struggled to learn to read and write, but he was fascinated by art. At the age of 14, Mind's curious images caught the attention of artist Sigmund Freudenberger, who employed the teenager to colour his drawings. Mind had meticulous attention to detail, and he would even correct Freudenberger's lines if he thought that any of them looked out of place.

Mind didn't enjoy much freedom while he was working for the artist, but when Freudenberger died in November 1801, his world suddenly opened up. Other artists tried to employ him, but Mind carried on living with Freudenberger's widow. He would venture out into the city of Bern during the day to watch animals and children play and return home at night to sit with his cats and work on his pictures.

SUSIE AND THE SAVANTS

THE WOMAN WHO CAN'T REMEMBER

MEET THE WOMAN PERPETUALLY LIVING IN THE MOMENT DUE TO AN UNUSUAL AND FASCINATING MEMORY CONDITION

Words by **Peter Fenech**

Memories are something the majority of us take for granted. The ability to recall details of places, people and events we have experienced over the course of our lives are things we regard as guaranteed results of those experiences. It is a safe assumption that most of us never stop to consider what it would be like to exist without those recollections. Our memories are part of who we are and help to shape our future selves.

For Susie McKinnon, however, detailed memories are a foreign concept – she has none, or at least not of the type the rest of us are accustomed to. McKinnon is the first person to be officially diagnosed with Severely Deficient Autobiographical Memory Syndrome (SDAM), a condition that prohibits her from remembering detailed information about her experiences. While she may know she was present during an event – her own wedding, for example – she is unable to remember key details about the day. What people were wearing, the condition of the weather, things people said and did, the order of proceedings, the smell of food and what dishes were on offer, the colour of her own shoes... these are all aspects that, for McKinnon, have long since faded from her memory, or perhaps were never recorded at all.

There are many recorded cases of people living with memory deficiencies caused by brain damage, suffered as a result of an accident or from incorrect formation before birth. Equally there are thousands of patients who experience a deterioration in memory due to dementia – a condition commonly seen in the elderly. McKinnon's condition is unique in that she has not lost memories over time, nor has she noted a sudden loss in a single event. She has never been able to recall high-resolution information, as seen from the first-person perspective, i.e. autobiographical memory.

For most people, memories manifest as a visual recreation of a scene, viewed through the person's own eyes, which in turn trigger recollections of olfactory stimuli or tastes and textures experienced in an associated context. For an

SUSIE McKINNON

A core advantage of reduced episodic memory in modern humans may be freedom from life-long phobias and irrational fears. This could result from dissociation between contextual triggers and emotional responses

"McKinnon's condition is unique in that she has not lost memories over time… she has never been able to recall high-resolution information"

individual with SDAM, these imprints are missing: in essence they possess no 'mind's eye', or what is scientifically referred to as autonoetic consciousness. This inhibits the person's ability to mentally revisit a past moment and observe it from the first-person perspective.

McKinnon first became aware that she remembered things differently to her peers when taking part in a memory test for a friend at school, who was training to be a physician's assistant. She was confused by questions regarding details of her childhood, of which she was clearly expected to possess in-depth memories. While her friend suggested she should investigate further by taking more memory tests, conducted by specialists, it wasn't until 2004 that McKinnon discovered an article about varying forms of memory. This piqued her interest as she drew parallels to herself.

As well as our memory being divided into short- and long-term processes, decades of research has shown that our memory stores are further compartmentalised. While there is still some uncertainty surrounding the boundaries between these, it is now widely accepted that longer-term storage of facts, concepts and experiences is discrete. Implicit memories are largely subconscious and incorporate learning of skills and core functional tasks, termed procedural memory. Explicit, conscious memories are those we can actively recall – facts about ourselves and our world (semantic memory), in addition to details of experiences and specific events in our lives (episodic memory).

While McKinnon and others with SDAM have trouble recalling fine details of specific events or places – they lack episodic storage – she is still a fully functional member of society, able to remember job training, language skills and with a high-performing spatial memory, meaning she can easily find her way, with a great sense of direction. This is due to her semantic memory being intact, giving her the unhindered ability to store learned information, obtained over a long period of time.

This explains how McKinnon is able to remember the lyrics to a song and to read music, enabling her

SUSIE AND THE SAVANTS

A reduced ability to recall visual information may be a contributing cause of SDAM but equally may be a consequence

"What sets SDAM apart from other brain diseases or defects is that there are no identifiable secondary side-effects"

to perform solo parts in a choir, yet find it impossible to remember actually making the performance. She is even unable to recall how such an event made her feel – when asked about this she can only assume she must have felt nervous to perform in front of a crowd but confident in her abilities. She is unable to draw together various snippets of observed sensory information to form an impression of a single event.

What sets SDAM apart from other brain diseases or defects is that there are no identifiable secondary side-effects. The article Susie McKinnon read was about the work of psychologist Endel Tulving, who first made the distinction between semantic and episodic memory. Tulving was arguing that there must be individuals who, apart from lacking episodic retention, would be living otherwise normal lives. It was this characteristic that made McKinnon believe she was one of the subjects Tulving was describing and that she was part of an unidentified group of people with an as-yet unclassified disorder.

The exact causes of SDAM are largely uncertain. This is due in part to the fact there have been relatively few confirmed cases to use as study references and because cause and effect paradoxes are yet to be addressed. The latter refers to the ambiguous relationship between suggested linked medical and psychological factors and the symptoms of SDAM – do these possible causes result in deficient memory or are they in themselves a result of SDAM?

In a study by Sheldon et al in 2016, the researchers noted that in people with strong episodic memory there was a great deal of interaction between different lobes of the brain, in particular the posterior regions and medial temporal lobes. This suggests that, to form an autobiographical memory, we pull together related information that is processed by discrete brain regions, such as images and smells. From this we can piece together not only what each component means but also when and where we recorded the details – an episode of life. There have

SUSIE MCKINNON

While slightly paradoxical, Susie shows little interest in keeping physical records of her life as an autobiographical memory substitute

also been suggested structural differences in the brains of SDAM-positive individuals. The hippocampi (sides of the brain) are asymmetrical in these people, with the right hippocampus being significantly smaller. However, it is difficult to say whether this could result in reduced autobiographical memory or if SDAM leads to underuse of these brain areas, thereby causing them to be underdeveloped.

For people with 'normal' autobiographical memory, McKinnon's condition seems like a nightmare. She is unable to share her husband's memories of their first date or any of the holidays they have taken together. She has many friends but cannot reminisce about times she has spent with them. This has the potential to limit the social possibilities of a person with SDAM. McKinnon herself has explained how most of her peers 'don't get it' – others assume she merely lacks attention to detail. SDAM sufferers may often have to explain that their condition is very different. However, McKinnon also highlights that she is able to live in the moment with greater ease, never dwelling on her past or future. There are also painful memories – such as an occasion where her husband was subject to a racially motivated physical attack – that do not have a continuous impact on her mental wellbeing.

As far as McKinnon is concerned, her SDAM has not significantly reduced her quality of life. She thinks of herself as happy. While continuous detail recording has advantages, in a world where external data storage is easier, there is an argument for episodic memory becoming less critical. Also, if like McKinnon you want to perform in the arts, forgetting how scared it makes you could be a catalyst for unhindered creative advancement.

An additional advantage of work on SDAM is that findings may benefit research on other brain disorders

THE PEOPLE WHO NEVER FORGET
IS REMEMBERING EVERY MOMENT OF YOUR LIFE A BLESSING OR A CURSE?

Another fascinating memory condition is known as hyperthymesia, which is essentially the inverse scenario to SDAM. Individuals with hyperthymesia never forget details of their lives, obtained through their autobiographical memory. Recall is near instant and is involuntary – with only a small prompt, affected subjects can remember the date a minor life event occurred, what the weather was like and such obscure facts as the exact time they got a call from a friend that day. Unlike savants or some people with autism, hyperthymesia-positive individuals are not calendrical calculators, able to identify dates decades in the future. The details they remember are usually personal and significant to them on an emotional level.

As with SDAM, the causes of hyperthymesia are difficult to isolate. Structurally, research has found stronger, more abundant white matter linking brain regions in those with the condition, while, psychologically speaking, hyperthymesia has been linked to obsessive compulsive behaviours regarding dates. Both of these may contribute to the condition but may also result from having it. In many ways hyperthymesia can prove more life-degrading than SDAM – it has proven to be quite disturbing to those positively diagnosed, with many developing an extreme obsession with documenting every life event.

People with hyperthymesia often obsessively document life details, despite possessing an extraordinary episodic memory

099

SUSIE AND THE SAVANTS

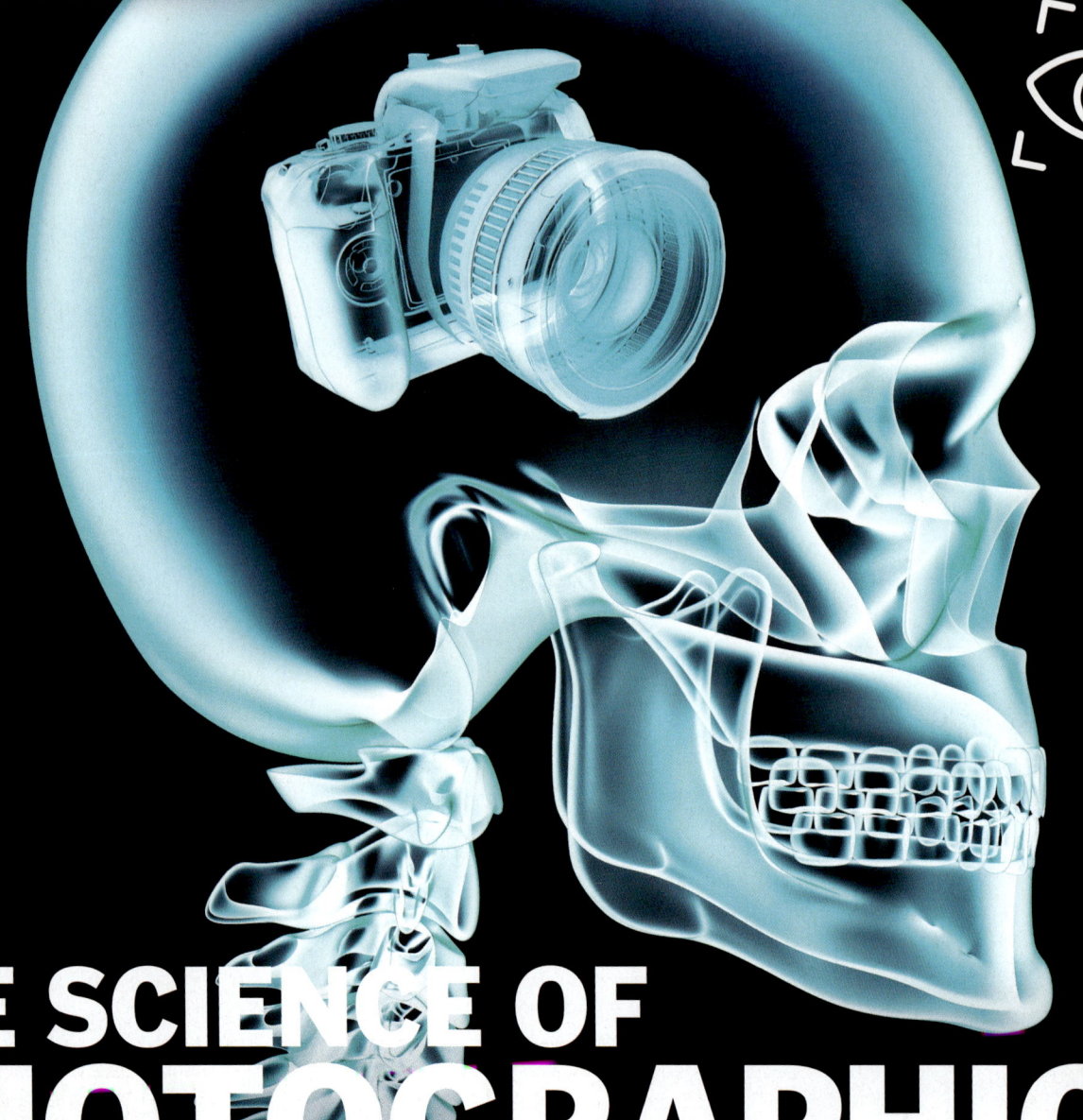

THE SCIENCE OF PHOTOGRAPHIC MEMORY

BUILDING PHOTO ALBUMS IN YOUR HEAD SOUNDS TOO IMPRESSIVE TO BE TRUE, BUT IS IT?

Words by **Ailsa Harvey**

We've all had that moment when we look at a photograph and our memories of that particular time come flooding back. But how much detail can you say you truly remember? What if you could remember every miniscule detail, from the paintings on the walls to the number of tiles on the floor – years later.

Photographic memory describes the ability to recall memories visually, almost as if referring to a photograph inside the head. While this is the common belief of what it means to have a photographic memory, evidence suggests this is not quite the case.

The majority of scientists believe that what many misinterpret as a photographic memory is actually eidetic memory. This is the ability of someone to remember an image in their mind for a period of time, even once the sight has been removed from their eyes. While some people see this image better than others and for a longer period of time, the image is not thought to be presented as a photograph in the head, which is something that can be scanned over to focus on different areas. Instead, the image is presented from one point of view and is less clear than a photograph would be.

People with eidetic skills are more likely to remember elements of a memory like a jigsaw. They are able to hold these puzzle pieces in their head for prolonged periods of time. The number of memories they acquire are not thought to be higher than the average person, but the percentage of memories moved into the long-term memory is. However, as can happen with jigsaws, pieces can go missing. It is typical for those with eidetic memory to forget some elements of the memory, but they have enough pieces of the picture saved in their mind to gather a pretty accurate interpretation of the whole scene.

PHOTOGRAPHIC MEMORY

Proof that people don't capture visual memories as a photograph in their heads comes from an experiment performed regularly with eidetic memory. When remembering a paragraph of text, someone who has snapped a photo would be able to read the words backwards if asked to do so. However, studies find that in this scenario, people who claim to have a photographic memory can't actually do this. They instead have to recall each part of the sentence as a separate memory.

It is important to note that we all have eidetic memory to a certain degree. The brain has the ability to remember what has been seen, but some people can refer back to this imagery for longer than others. To test your eidetic memory, all you have to do is look at the scene in front of you, shut your eyes and then see how long you can remember the view in significant detail. For most, the vision starts to fade after a few seconds.

When the average person observes something, such as a messy bedroom, they will leave their house with the vision of this untidy room that they need to sort. Moving through the day, this will be put to the back of their mind, but when the thought of their bedroom comes back, they are likely to still remember that they need to tidy it. The difference is that this has become information in their head at this point: they are not viewing a photographic image when they have this memory.

So what is it like to be able to hold onto this picture for a matter of days, months or maybe years? Is this even possible?

If it exists, photographic memory continues to stun those who don't have the ability to comprehend it. Many see it as a superpower, and some look on in awe and jealousy. However, an intensely powerful memory can sometimes be more challenging than it sounds. In some cases, people storing unnecessary memories have their heads filled. Times that they don't need to remember circle around in their heads.

There are fewer than 100 people known worldwide who have a condition called highly superior autographical memory (HSAM). This is such a refined memory that it can be incorrectly mistaken for a photographic memory. People with this kind of memory can remember events in great detail from decades ago, such as what day of the week something happened, what they were wearing on any day and what meals they and others ate.

Undoubtedly, there are people with expert memory among us, but whether there are truly any minds capable of holding memories in the form of photographs remains unknown. Unfortunately, there is not yet a way to prove exactly what is being seen in the mind's eye of others. Those who claim to be able to use photographic memory are the only ones who can know what is going on in their head, but there is much speculation in the scientific world.

> "The brain has the ability to remember what has been seen, but some can refer back to this imagery for longer than others"

Between two and 15 per cent of children can memorise vivid images for a few minutes or more after closing their eyes

CHARLES AND ELIZABETH
TAKE A LOOK AT THE MOST CONVINCING PHOTOGRAPHIC MEMORY TO DATE

One experiment sits above the rest in exploring photographic memory potential. In 1970, scientist Charles Stromeyer studied the memory skills of student Elizabeth. In the experiment, he covered her right eye and showed her left a series of 10,000 dots. Then the following day he did the same with a different series of dots on her right eye.

Amazingly, Elizabeth was able to recall the position of the dots and recreate a three-dimensional structure using the correct series of 20,000 dots. In order to successfully achieve this, she had to refer back to the two images she had seen in her head, and recall the dots' positions to perfection. While this fits the definition of a photographic memory, it could only be assumed what was happening in her brain for her to remember this. As no evidence has been as compelling since, is this the closest we will come to finding a photographic mind?

Elizabeth was a student at Harvard University

SUSIE AND THE SAVANTS

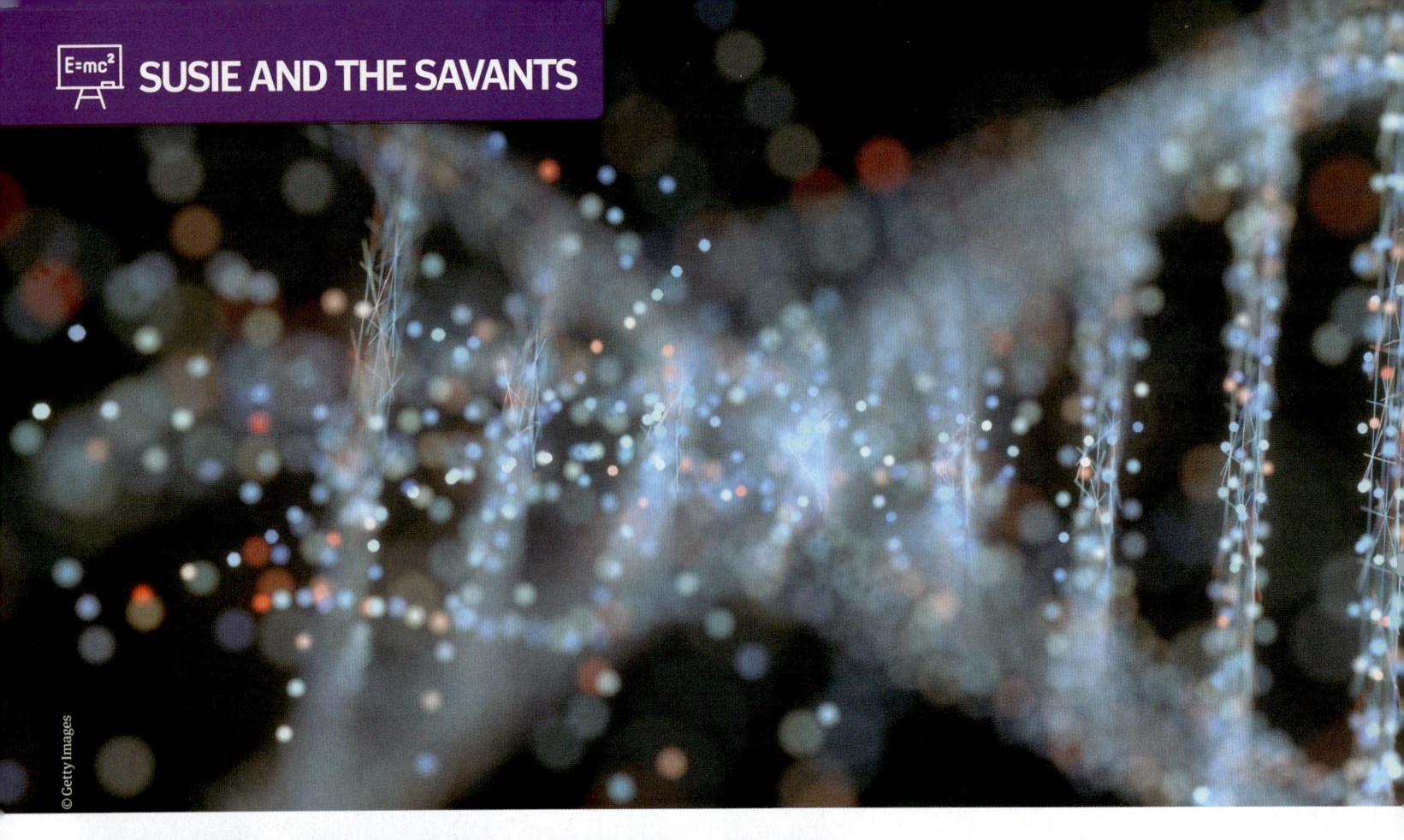

THE SCIENCE OF GENERATIONAL MEMORY

DO WE KNOW THINGS BEFORE WE'RE BORN BASED ON THE EXPERIENCES OF THOSE BEFORE US?

Words by **Nikole Robinson**

Your genes make you who you are. They're inherited from your parents, and theirs from their parents before them, going all the way back to the time humans first emerged. Along the way they've picked up traits that help us survive as a species, diversifying for humans living in different parts of the world with distinct needs and evolving along with the environments around us.

Much of this is encoded as natural instinct. When a baby is born, it doesn't need to be taught how to breathe or cry when it needs attention – this is hard-wired into our DNA. We know that when we're tired we need to sleep and when we're hungry we need to eat. The same goes for animals. Within two hours of being born a foal attempts its first steps with little guidance from its parent, all from instinct.

As well as our basic instincts, our genetics can carry more complex abilities, knowledge and conditioning. One example of inherited behaviour exhibits itself through reflexes that keep us safe in situations where we perceive danger, such as pulling our hand back when it meets something hot or jumping when we're startled.

Findings have also hinted that our emotional responses and facial expressions could be hard-wired behaviours. Babies start to smile and laugh at around two and four months respectively – again this isn't something they're ever specifically taught. Although they may take in the expressions and moods of those they grow up around, it's hard to test if they connect emotion to the faces they see. However, a 2009 study showed that athletes who had been born blind made the same disheartened expressions when losing as athletes with sight, suggesting that facial expressions may not be something we learn from watching others.

While genetics play a part, infants are also heavily influenced during their time in the womb. Mothers who eat strong-tasting foods such as garlic can affect the taste buds of their children, given them an affinity for the same tastes later in life. Taste and smell are important triggers that babies look for when first

"Inherited behaviour exhibits itself through reflexes that keep us safe, such as jumping when we're startled"

GENERATIONAL MEMORY

It's not just visual traits that are passed down from parent to child

feeding, as the mother's milk will carry the same flavours experienced in the womb. This pre-birth taste sampling may have developed as a way to teach newborns to feed, increasing their chances of survival. This prenatal flavour learning is also seen in other mammals such as rabbits, cats and dogs.

Whether we have a predisposition to a certain language is still something that's contested – including whether our parents' native tongue comes more naturally to us through genetics or merely from exposure. However, what is clear is that humans have a biological preparedness for learning a language, often speaking in full sentences by the age of four. Again, prenatal learning could influence this, as it's been proven that a baby can hear sounds (the part of the brain that processes these noises can become active in the last trimester).

Looking to the animal kingdom, fairy wrens also 'speak' to their eggs in this way, with the mother singing a specific song that her chicks will later chirp back to her once hatched. This is used as a way to identify each other, almost like a mother-child password. However, suboscine birds seem to have the song of their species ingrained. Chicks hatched from eggs incubated in soundproof solitude are able to produce the correct call for their species having never heard it, meaning this could be true for humans too.

Generational memory could also help explain why certain people are naturally gifted in areas such as music, sport or mathematics. It's argued that they could be unlocking knowledge stored from ancestry, tapping into the talent of generations ago. The problem is that the existence of generational memory is hard to prove, since there are so many environmental and outside factors that change behaviour and personality, especially with young minds, which are so easily moulded. However, the theory is starting to be taken more seriously, and studies involving short-generation life forms such as nematode worms are making significant breakthroughs. We may soon have undeniable proof that our genetics pass down more than we think.

INHERENT ANIMAL BEHAVIOURS

HOW FEAR WAS SHOWN TO BE PASSED DOWN THROUGH GENERATIONS OF MICE

In a 2013 study to explore if phobias and anxiety can be transferred genetically, neuroscientists from Emory University in Atlanta, Georgia, trained male mice to fear the scent of cherry blossoms by associating it with a mild shock. These mice were then bred, and their offspring were raised without ever being exposed to the smell. Upon reaching adulthood the second-generation mice were introduced to the same smell. Despite never experiencing the shock themselves, the scent alone was enough for them to exhibit signs of anxiety and fear. It was also discovered that their noses had more neurons specifically designed to detect this scent, as well as more brain space dedicated to recognising it. So this wasn't just a behavioural change – it was physical too.

Even more surprising was that when these mice bred, the grandchildren of the original mice exhibited a similar reaction to the scent. The experiment confirmed that genetic markers could be used to transmit a traumatic experience to future generations. We acquire these epigenetic markers throughout our life by learning and interacting with our environment. By teaching the next generation what to be fearful of without them experiencing the trauma themselves could help the survival of a species.

Mice have amazing memories and are known never to forget a route once they've memorised it

SUSIE AND THE SAVANTS

MEMORIES OF A PAST LIFE

AROUND THE WORLD, THOUSANDS OF CHILDREN REMEMBER A LIFE THEY LIVED YEARS AGO

Words by **Baljeet Panesar**

Over a period of 40 years, Professor Ian Stevenson compiled roughly 3,000 cases of children who had memories of past lives from around the world. He authored over 300 publications and travelled nearly 90,000 kilometres to document these cases.

In instances of reincarnation, most children start talking about a past life as soon as they can speak, between the ages of two and three. These memories tend to fade between seven and nine. Each child experiences vivid and remarkably accurate descriptions of their previous lives. Most of these cases tend to occur in south Asia and western Asia – cultures in which there is a strong belief in reincarnation – but there have also been cases in Western Europe and North America. Stevenson published the cases of 73 American children who had past lives in 1983, and 21 European cases, nine of which were from the UK, in 2003. His attention to detail, use of birth certificates and post-mortem results and interviews with witnesses to confirm a child's description of their past life were characteristic of his diligent and tenacious approach.

Stevenson's masterpiece was the two-volume, 2,268-page monograph *Reincarnation and Biology: A Contribution to the Etiology of Birthmarks and Birth Defects* (1997). In this book he reports detailed studies of more than 200 children who seem to have visible birthmarks and birth defects that can be traced to the past life of the children. These wounds had occurred during their previous death – often violent and sudden – which the children remember. Unlike the small and flat birthmarks most of us have, Stevenson observed that these children had raised or puckered, unusually shaped marks, possibly even two – one that would represent an entry wound and the other the exit wound. Stevenson noted 18 such cases.

As a result of their past lives, these children often experience confusion. Children may fear the police because they were involved in their previous death, or they may be wary of certain weapons that were used to kill them. These phobias seem irrational because to their parents' knowledge their child has not had any traumatic experiences. These children may also long for their favourite clothes, food or stuffed animal – behaviours associated with their previous lives. Stevenson's work is continued by child psychiatrist Jim Tucker.

> *"Children may fear the police because they were involved in their previous death, or they may be wary of certain weapons"*

PAST LIVES

THE BOY WHOSE REBIRTH WAS PREDICTED BY HIS GRANDFATHER

Before his death, William George Sr., a fisherman from the Tlingit Indian tribe of Alaska, told his son that he would return as his baby boy and that they would recognise him by the two half-inch birthmarks on his body. Nine months after George Sr. had been lost at sea, baby William George Jr. was born on 5 May 1950. The baby had two very distinctive birthmarks in the same location as his grandfather. As the boy grew, those closest to him noticed how similar the boy was to his grandfather; he looked like him, walked like him and had developed knowledge about fishing and boats that you wouldn't expect from a child so young. But he, unlike other children his age, feared the water. Aged four, he noticed a gold watch that had belonged to George Sr. and insisted that it was his. In later childhood the boy did forget his past life experiences, but he continued to lay claim to the watch.

THE BOY WHO WAS MURDERED BY HIS OWN BROTHER

On 26 August 1956, the Gupta family welcomed their baby boy, Gopal, into the world. At the age of two an incident occurred at the family home. He refused to clear a glass from the table, to the astonishment of his parents; he had no need to work, he said, because they have servants to do the household chores. In his previous life he had lived in a city called Mathura, 257 kilometres from his home in Delhi. Here, he had two brothers – one of which shot him – and another father. He claimed that he owned a pharmaceutical company called Sukh Shancharak. Eight years later it emerged that details of Gopal's life – and death – were similar to that of Shakipal Sharma, who had been shot by his younger brother in 1948. He was able to find his way to Shakipal's house and to the Sukh Shancharak company, as well as pointing out where and who shot him. He could also identify people in photographs at the Sharma household.

© Getty Images

SUSIE AND THE SAVANTS

THE GIRL WHO DIED DURING CHILDBIRTH

In 1926, a baby girl named Shanti Devi was born in Delhi, India. At the age of three she started talking about a former life in Muttra (now known as Mathura), a town 130 kilometres away. She said that she was born in 1902, that her name had been Lugdi, and she had been married to a cloth merchant by the name of Kedar Nath Chaubey. She also said that she had given birth to their fourth child and had died ten days later.

When Shanti was nine years old her family decided to find out whether their daughter's statements were correct. They were. A year later she travelled to Muttra, identifying relatives of Kedar Nath at the train station. She was also able to direct witnesses to her former husband's house. She even recognised her former parents – and knew their names – in a crowd of 50 people, and she knew Muttri idioms and spoke the dialect. During her time in Muttra she made at least 24 statements, all of which were verified.

THE BOY WHO ATE FAR TOO MUCH CURD

Born in 1944 in Bisauli, India, two-and-a-half-year-old Parmod Sharma would tell his mother that his wife in Moradabad, a city 145 kilometres away, would cook for them. Then, between the ages of three and four, he would tell of the soda and biscuit shop he once owned in Moradabad called 'Mohan Brothers'. He said that he once became ill by eating too much curd and that he died in a bathtub. The boy's statements reached the Mehra family in Moradabad, who owned a soda and biscuit company called Mohan Brothers. Their brother and partner Parmanand Mehra had died in 1943 after contracting a fatal illness caused by eating too much curd. At the age of five Parmod visited Moradabad, recognising some members of the Mehra family and places around the city centre. Later, Parmod travelled to Saharanpur, where the Mehra family also owned businesses, and recognised people in the city.

PAST LIVES

THE BOY WHO MURDERED HIS WIFE

H. A. Wijeratne Hami was born in the village of Uggalkaltota, Ceylon (now known as Sri Lanka), in 1947. He was born with a visible deformity to his arm and right breast – a symptom of Poland syndrome. When the boy was two years old, his mother overheard him saying that his arm was deformed because he had murdered his wife, Podi Menike, in a former life. She mentioned this to her husband, who told her that his younger brother, Ratran Hami, had been executed for murdering his wife in 1928. Before he died, Ratran had told his older brother (Wijeratne's father) that he would return. Wijeratne was able to recall the details of the crime, arrest and execution of Ratran, which were confirmed by court documents and interviews. Wijeratne regarded the deformity as a punishment for the crime that he, as Ratran, committed.

THE GIRL WHO WAS A BOY

Ruby Kusuma Silva was born in 1962 in the Sri Lankan city of Galle and started speaking about her previous life before she was two years old. Her former life was in Aluthwala, about 14 kilometres from where she lived. She said she had two brothers, that her father was a bus driver, and there were lots of coconut trees, speaking as if she was still living that life. She seemed confused about no longer being a boy. She said that she had drowned after falling into a well (she had phobia of wells in her present life until the age of eight). After some investigation, it was found that the Singho family had a son named Karunasena who had drowned in a well aged seven in 1959. Ruby herself had displayed more male-associated behaviours, like wanting to wear boy's clothing, asking her family to call her brother or son (not sister or daughter) and climbing trees.

SUSIE AND THE SAVANTS

THE BOY WHO REMEMBERED HIS PAST LIFE AFTER HE NEARLY DIED

In 1954, three-year-old Jasbir Lal Jat was thought to have died of smallpox, but a day after he appeared to revive. It took a number of days for him to be able to speak again, but he was no longer the same person his family had brought up. He now believed that he was the son of a Shankar from the village of Vehedi and wanted to go there. For almost two years he refused to eat with his family, saying he belonged to the Brahmin caste – a higher caste than the Jats. He spoke of how a man who owed him money gave him some poisoned sweets that caused him to fall off his chariot, inflicting a fatal head injury. When word reached the Tyagi family, they realised that Jasbir's life and death were similar to that of their son, 22-year-old Sobha Ram, who had passed away earlier in the year. While in Vehedi, Jasbir was able to find his way around the village, recognise members of the Tyagi family and demonstrate the workings of the family and their affairs. This type of 'replacement reincarnation' is much rarer than other types of reincarnation.

THE REBORN WWII FIGHTER PILOT

James Madison Leininger was born to Protestant parents Bruce and Andrea Leininger in San Francisco in 1998. When he was 22 months old he started developing a fascination with aeroplanes – notably World War II aircraft – after visiting the James Cavanaugh Flight Museum in Texas. Shortly after he started having nightmares, recalling the incident of his past life. From the age of 28 months his recollections became more detailed: his plane had crashed after being shot down by Japanese forces near Iwo Jima; he could fly a Corsair; and he knew the name of his copilot, Jack Larson. But his identity in his past life was not that of Larson (he was still alive), but rather that of James M. Huston Jr., who had been killed aged 21 and whose death appeared to match James' statements.

PAST LIVES

A CHRISTIAN BOY BORN AGAIN AS A BUDDHIST

Gamini Jayasena was born in Colombo, Sri Lanka, in 1962. When he was a year and a half old he started talking about his former life, saying that his other mother was his 'real mother'. He mentioned someone called Nimal, who had bitten Gamini, and that Nimal had also been bitten by a dog. The Jayasena family were Buddhists, but Gamini prayed in the position of a Christian, as well as celebrating Christmas and speaking about Santa Claus. His family suspected that in Gamini's past life he had been a Christian. While on a family bus trip when Gamini was two and a half, Gamini's relatives stopped at a town called Nittambuwa, 35 kilometres from their home, and Gamini immediately recognised it as the home of his past life. Here, an eight-year-old Palitha Senewiratne had died some years before after developing a serious illness. During later visits, he recalled how his past life father would break off olive branches for him, selected his favourite past life sweet and found his past life boarding house at his school.

PROFESSOR IAN STEVENSON
THE MEDICAL DOCTOR WHO TURNED TO THE STUDY OF REINCARNATION

Born in Montreal, Canada, on 31 October 1918, Professor Ian Stevenson was educated at St Andrews University, Scotland, and McGill University, Canada. He received his medical degree from McGill in 1943 and then spent some time working in biochemical research.

In the late 1940s he started working at New York Hospital and began research in psychosomatic medicine, in particular the effects of stress on physical symptoms. In 1957 he was appointed head of the department of psychiatry at the University of Virginia. Inspired by a meeting with Aldous Huxley, Stevenson became one of the first academics to research the effects of psychedelic drugs in the US.

After publishing his work in 1960, Chester Carlson, the investor of the Xerox machine, funded a trip to India and Sri Lanka, where Stevenson completed further research. Seven years later he would become the first Carlson Professor of Psychiatry.

At the age of 88, Stevenson passed away on 8 February 2007 from pneumonia. He will always be remembered as one of the greatest authorities on reincarnation, near-death experiences and the paranormal.

THE MAN WHOSE MURDERERS HAD GOT AWAY WITH IT

In 1962, Bongkuch Promsin was born near the town of Tha Tako, Thailand, but at the age of 20 months he started speaking about his previous life, saying that his current home wasn't his actual home. He usually said this after waking up.

At the age of two he said that his name had been Chamrat, that he was a Loatian (a member of a tribe native to Southeast Asia), had owned cows and had been murdered by two men after attending a fair in Hua Thanon, six kilometres from his family home.

Eventually, word of Bongkuch's testimony reached the family of Chamrat Pooh Kio, who had been murdered at the age of 18. The Promsins visited Chamrat's family twice in 1964, where they confirmed most of Bongkuch's statements. Unlike most children, Bongkuch didn't spontaneously speak about his previous life often, mostly doing so when something reminded him of his past life. By the age of 12 he no longer spoke about it.

FLORA, FAUNA AND THE FUTURE

112 ANIMAL MEMORY

122 AI AND THE FUTURE OF MEMORY

FLORA, FAUNA AND THE FUTURE

120 PLANT MEMORY

FLORA, FAUNA AND THE FUTURE

ANIMALS WITH EXCELLENT MEMORIES

FROM RESENTFUL BIRDS TO LOGICAL SEA LIONS, MEET THE ANIMALS PUTTING OUR OWN MEMORIES TO SHAME

Words by **Victoria Williams**

For a long time it was believed that humans were the only beings that were capable of exhibiting 'advanced' cognitive processes like deduction and remembering information. However, in the short time that scientists have been studying the memories of other species, there have been a lot of surprises. Animals ranging from apes to insects have demonstrated unexpected abilities and applications.

A recent study of 25 species (including dogs, bees, rats and dolphins) found that the average short-term memory of an animal lasts 27 seconds. Dogs lose all recollection of an event two minutes after it happens, while chimps can only hold on to memories for about 20 seconds. Short-term memory appears to be one area where humans are unusually strong, but when it comes to long-term memory, the playing field is a little more even.

While most animals are unable to remember specific events more than a few seconds afterwards, many are capable of creating stores of information that they might need later. This relies on 'associative memory'; a dog stung by a bee will soon forget the trauma of the actual attack, but connections that are subsequently formed in its brain mean that it will associate bees' nests with danger in the future and thus stay well away. Experts think that animals may have highly specialised systems in their brains that make sure 'biologically relevant' information like food sources and threats are stored as long-term memories.

Compared to our knowledge of other areas of their lives, scientists are only in the early stages of understanding animal memory. There may be species out there with memories stronger than anyone would expect. After hundreds of years of being underestimated, we're just beginning to realise that animals might know a lot more than we think they do...

ANIMALS WITH EXCELLENT MEMORIES

DOLPHINS

Dolphins are always among the first to come up in discussions of animal intelligence, and they appear to have long-term memories to match. Research into bottlenose dolphins has shown that each one has a 'name' – a unique whistle used by the other members of a pod. A recent study of captive dolphins that had moved between different facilities found that they recognised and responded to the whistles of old acquaintances even after 20 years apart.

In the wild, bottlenose dolphins live in a fission-fusion society; the size and composition of each pod changes frequently as groups split, join up again and form new combinations. Remembering past pod-mates and maintaining relationships over many years might make the complex task of choosing and slotting into a new group slightly smoother for these highly social creatures.

FLORA, FAUNA AND THE FUTURE

ELEPHANTS

They say an elephant never forgets, and there's some truth to the saying. While they can't remember everything they've ever experienced, elephants do have an incredibly strong ability to store and recall important information.

Living in groups that range over large territories, social and spatial memory are both vital for survival. Matriarchs, the leaders and decision-makers of the herds, have shown especially good memories for relationships with other elephants and routes to food and water sources. Distinguishing relatives from strangers helps keep vulnerable calves safe from potential conflict, and knowledge of the area could be life-saving if a foraging site or watering hole became inaccessible.

According to research, matriarchs are capable of remembering both the identities and the last known locations of up to 30 other elephants, creating a mental map that allows the herd to regroup after breaking up into smaller foraging parties.

CLARK'S NUTCRACKERS

Corvids are known for their intelligence, and Clark's nutcracker certainly lives up to the family reputation. Using a pouch under its tongue that can hold up to 150 pine seeds at a time, this industrious bird flies through the forest collecting food and storing it in the ground for the cold months. Seeds are usually cached in groups of three or four, and over a season a Clark's nutcracker can stash away more than 90,000.

Nine months later it demonstrates its strong long-term spatial memory by recalling the locations of around 20,000 of its hiding spots and retrieving the seeds, even when the ground is covered in thick snow. Remembering as many cache sites as possible is important for this bird, but it's the few that get forgotten that benefit the forest; left in the ground, some of the seeds germinate and begin to grow into new pines.

ANIMALS WITH EXCELLENT MEMORIES

BUTTERFLIES AND MOTHS

Moths and butterflies might not have vast memories like some of the other species on this list, but there is something quite remarkable about them: they can remember things they learned as caterpillars. Despite most of their tissues breaking down and reforming during metamorphosis, scientists have discovered that at least part of the brain is preserved. Researchers used mild electric shocks to teach tobacco hornworm caterpillars to avoid ethyl acetate, a chemical used in nail varnish remover, and found that almost 80 per cent still avoided the scent after transforming into adult moths.

FLORA, FAUNA AND THE FUTURE

HORSES

As herd animals, understanding social cues and relationships are vital skills for horses. This ability goes beyond members of their own species: horses have been shown to read and remember the facial expressions of humans too.

Scientists showed horses a photo of a human model looking either happy or angry. When the model arrived in person with a neutral expression, horses that had seen them scowling in the photo became more agitated. Those that had seen a smiling photo spent more time looking at the model with their right eye, the preferred eye for looking at social stimuli and positive sights.

PAPER WASPS

The possibility of intelligence in insects was dismissed for a long time because it was thought that their brains were just too small to manage more than basic functions, but wasps have proven to be surprisingly good with faces.

Paper wasps are 'eusocial', meaning that they live in large, structured groups that cooperate to raise young and defend their homes. Experts already knew that these wasps could recognise the face patterns of individuals they were in frequent contact with – an ability that probably makes it easier to detect intruders – so the next logical step was to find out how long they could hold these patterns in their memories for. Michael Sheehan and Elizabeth Tibbetts of the University of Michigan decided to find out.

In a feat not seen before in any insect species, paper wasps were able to tell the difference between familiar and unfamiliar faces, even when they hadn't seen their nest mates for a week.

ANIMALS WITH EXCELLENT MEMORIES

RIO

After her mother didn't appear to be caring for her properly, Rio the Californian sea lion was raised by humans at the Long Marine Lab in California. Marine biologists began to train Rio when she was around six years old and were impressed by her ability to learn tricks more complicated than those usually found in sea lion displays. Rio began to show an understanding of 'sameness', choosing symbols matching those held up by the biologists to earn fish rewards.

In 2001, ten years after she'd been taught her matching trick, the researchers returned. Even though she hadn't performed the trick or been reminded of it since 1991, Rio was just as good at matching as she had been the first time around. This extraordinary sea lion was even able to match symbols she'd never seen before, demonstrating an understanding of the rules of the game rather than just a recollection of familiar symbols.

AYUMU

Ayumu the chimpanzee is a third-generation memory champ, part of a decades-long study into chimp cognition at the Primate Research Institute of Kyoto University in Japan.

Not long after his birth in 2000, Ayumu began to participate in memory tasks. His most famous skill involves remembering the position of nine randomly placed numbers that flash up on a touch-screen computer for a fraction of a second and then touching the positions in sequential order. Ayumu had quicker response times and higher accuracy rates than university students attempting the task, challenging the long-held belief that humans have an unrivalled working memory.

CROWS

With their bright beady eyes and history of being linked to all things dark and foreboding, it might not be a complete surprise to learn that crows use some of their formidable intelligence to judge and torment humans.

To be fair to these misrepresented birds, they only hold grudges against people who have treated them badly or caused them distress. Researchers confirmed this corvid justice system by wearing masks, capturing crows and banding their legs. Anyone later walking through the area with the same mask on was loudly 'scolded' by the banded crows, even three years after the birds had been caught.

It seems that the crows had focused on faces, because the dress, body and gait of the person wearing the mask didn't have to match the original researcher for them to begin their tirade.

FLORA, FAUNA AND THE FUTURE

CHASER

Chaser the border collie holds the record for the largest tested memory of a non-human animal. At just a few months old, Chaser started to show signs of understanding that certain objects had names, and from then on she worked with her owner – researcher John Pilley – to see just how much she was capable of learning.

By the time she died in 2019, Chaser could correctly identify and retrieve 1,022 toys by name. In addition to her impressive memory, Chaser was also capable of inference; on hearing an unfamiliar name she was able to work out that she must be looking for a new toy.

OCTOPUSES

The cephalopods – octopuses, squid and cuttlefish – are in a league of their own when it comes to invertebrate brain power. With much more complex nervous systems, their cognitive ability is thought to be nearer to that of a mammal than any of their close relatives. In fact, the common octopus' 500 million neurons comes close to the number found in the common marmoset.

Octopuses play, learn, solve problems and possibly even dream. They can recognise individual human faces and appear to remember them too. In 2010, biologists at the Seattle Aquarium presented Pacific octopuses with a 'nice' keeper and a 'nasty' keeper. The nice keeper fed the octopuses regularly, while the nasty keeper touched them with a stick tipped with bristles. After just two weeks of this treatment all eight octopuses behaved differently when the mean keeper entered the room.

ANIMALS WITH EXCELLENT MEMORIES

BLUE WHALES

Every year, blue whales set off on the long migration between their breeding and feeding grounds. Many migratory species seek out the best food sources available each season, but blue whales appear to be guided more by the past than the present. Combining ten years of data on the migratory movements of these whales, researchers observed with surprise that they timed their journeys and chose their routes to match the historical average timing and location of the annual boom in krill numbers.

Rather than swimming in search of their food as it becomes available, blue whales head for patches of the ocean where they remember krill being plentiful in previous years – choosing stable, high-quality feeding patches over new but potentially fleeting areas of abundant food.

A MIRACLE MEMORY BOOST?
SADLY NOT, BUT THAT ISN'T STOPPING PEOPLE BUYING PREVAGEN

You won't find jellyfish on this list because the gelatinous creatures have no brains – just a simple set of nerves that allows them to sense and react to their surroundings. Despite this, a protein found in jellyfish is being marketed as a miracle memory booster.

Prevagen, a 'supplement' that claims to improve memory and sharpen the mind, uses apoaequorin – a protein produced by *Aequorea victoria*, a bioluminescent species also known as the crystal jelly – as its main ingredient. There's no evidence that it has any effect on the human brain – the only research paper suggesting it might was written by the company behind Prevagen, and their methods weren't exactly standard.

Even after lawsuits were filed against the company, consumers continue to spend more than $60 (£50) for a month's supply. They're all dreaming of better memories, but taking the supplement will do about as much for their brains as eating a bowl of jelly.

The crystal jelly epitomises the saying, 'The lights are on, but no one's home'...

FLORA, FAUNA AND THE FUTURE

The daffodil is a prime example of plant memory, blooming according to the temperature and length of day

DO PLANTS HAVE A MEMORY?

WE ALL KNOW THAT MOTHER NATURE IS MESMERISING, BUT CAN ONE OF HER CHAMPIONED CREATIONS ACCESS THE CAPACITY TO LEARN AND FORM MEMORIES?

Words by **Laurie Newman**

To the untrained eye, it may look as though plants don't have any long-term plans, but in fact the opposite is true. Drooping when they lack water, growing towards sunlight and perking up through rain – all of these common responses to changes in environment are shared among many species and show that they retain information.

The *Mimosa pudica*, also known as the 'sensitive' or 'sleepy' plant, is one of the first that springs to mind when discussing whether plants have memory. The plant itself, aptly named 'pudica' from the Latin meaning 'shy', responds to gentle touch and stimulation by instinctively closing its leaves and drooping. The Venus flytrap (*Dionaea muscipula*) responds to touch in a similar way by ensnaring any prey that dices with death around its leaves. The latter remembers routinely that two of the hairs on its leaves need to be touched by a bug before it can shut.

Both of these plants' responses to touch suggest that memory is present, as they repeat the same pattern of behaviour relentlessly. Many scientists have studied the effects of memory on plants and concluded that there is more than one type present: short-term memory, immune memory and transgenerational memory. For example, while the short-term memory in a Venus flytrap is generated by electrical signals sent when a bug lands on its leaves, the longer-term memories are formed in the DNA and are passed on to other generations of the plant in transgenerational memory. The same can be said for seasonal plants like daffodils, which only bloom during certain months.

This is something commonly referred to as vernalisation, which Russian scientist Trofim Lysenko developed in 1928 following a famine in several regions of the Soviet Union. He observed that he could make cold weather plants flourish all year round by providing low temperatures for seeds and infant plants. Not only was this breakthrough ideal for

DO PLANTS HAVE A MEMORY?

agriculture, as it meant that crop yields could be manipulated to produce ample food, but it also showed that plants were remembering through temperature when to bloom.

In 1948, scientists Georg Melchers and Anton Lang continued important investigations into the mystery of plant memory through studying the henbane flower. They probed the limits of the plant's memory of winter by exposing it to short bursts of cold and hot temperatures. They found that the plants formed lasting impressions of the cold rather quickly. Although at the time not referred to as a study specifically looking into plant memory, it is now seen as one of the most influential breakthroughs. The experiments concluded that plants can hold on to their pasts for much longer than originally thought. Researchers are hoping that this evidence will one day enable us to 'train' plants for our benefit, making severe conditions like drought and small crop yields one less thing for farmers to worry about.

"They probed the limits of the plant's memory of winter by exposing it to short bursts of cold and hot temperatures"

Through memory and electrical signalling, the carnivorous Venus flytrap knows exactly when to shut its leaves

MIMOSA PUDICA
THE MIMOSA PUDICA HABITUALLY SHUTS ITS LEAVES THROUGH TOUCH, BUT AN AUSTRALIAN SCIENTIST HAS PROVEN THEY CAN BE TAUGHT NOT TO

The *Mimosa pudica* responds almost immediately to touch by dropping and folding its leaves inwards

In 2014, Australian biologist Monica Gagliano showed that plants may have memory and the capacity to learn by closely studying the *Mimosa pudica*, or the 'shy plant', which folds its leaves after being touched. Gagliano routinely dropped 56 of the plants from a height of six inches on more than 60 occasions to observe whether or not she could train them to not instinctively curl their leaves upon falling. After a period of time, Gagliano gradually noticed that the plants stopped closing their leaves upon impact, observing that the plants seemed to figure out that falling wasn't going to hurt them. A week later she resumed the dropping of the plants, and still they remained unfazed. Gagliano repeated this for a period of 28 days, and still the plants 'remembered' what they had learnt. Gagliano concluded in her paper that "plants may lack brains, but they do possess a sophisticated signalling network".

Trofim Lysenko made great discoveries into vernalisation. He specifically identified how plants use stored memories to know when to bloom annually

FLORA, FAUNA AND THE FUTURE

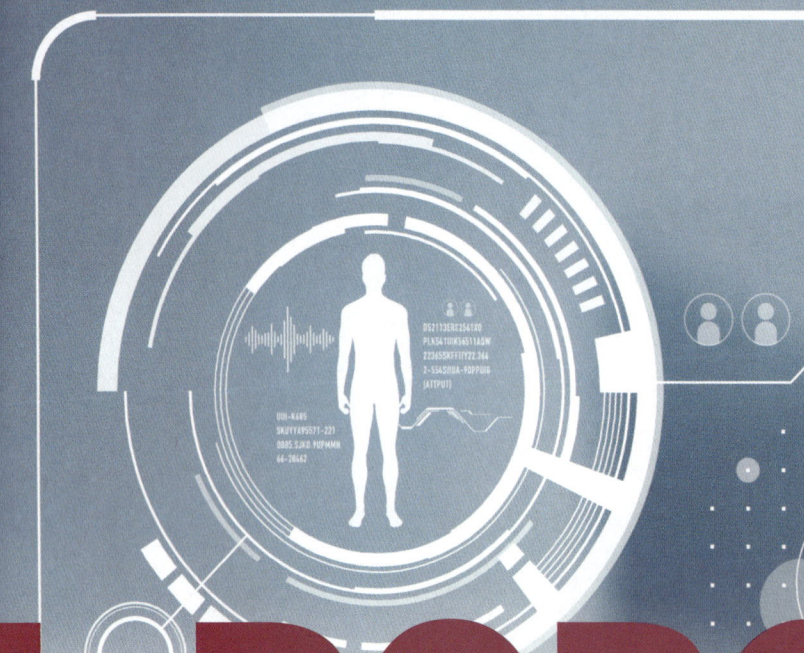

AI, ROBOTS AND THE FUTURE OF MEMORY

HOW COMPUTER SCIENTISTS ARE TAKING INSPIRATION FROM THE HUMAN BRAIN TO DESIGN INCREDIBLE VIRTUAL COPYCATS

Words by **James Horton**

Memory is possibly humanity's greatest tool. Without memory there would be no language, no advanced social structure – no plans at all. Memory is an exponential gift; the more we already know, the easier it often is to deconstruct and make sense of new but similar information. Our memories allow us to connect seemingly disparate parts of information together: an accomplished mathematician will transition into the field of physics with relative ease; a professional athlete will often learn a new sport much faster than an amateur. With such a powerful asset in our arsenal, it's understandable that so many computer scientists long to grant a similar gift to their software.

We understand a great deal about our own biological memory machinery, but our familiarity remains peppered with caveats and missing knowledge. At their core, memories are formed by neurons – brain cells that are entangled in a complex web of connectivity 100 trillion connections strong. When engaged, neurons fire and send a cascading electrical signal through the brain, stimulating their neighbours if the signal is strong enough. This much is understood, yet modern neuroscientific research shows indirect evidence of neuron-firing 'echoes', which continue to fire after the signal has ended. There is also evidence that during the formation of 'silent' memories – which are formed without our active, conscious participation – the activity of the brain is peculiar in its quietness. Computer scientists attempting to emulate the memory feats of the brain, then, are faced with two formidable obstacles. They must design a complex architecture that can rival the still incompletely

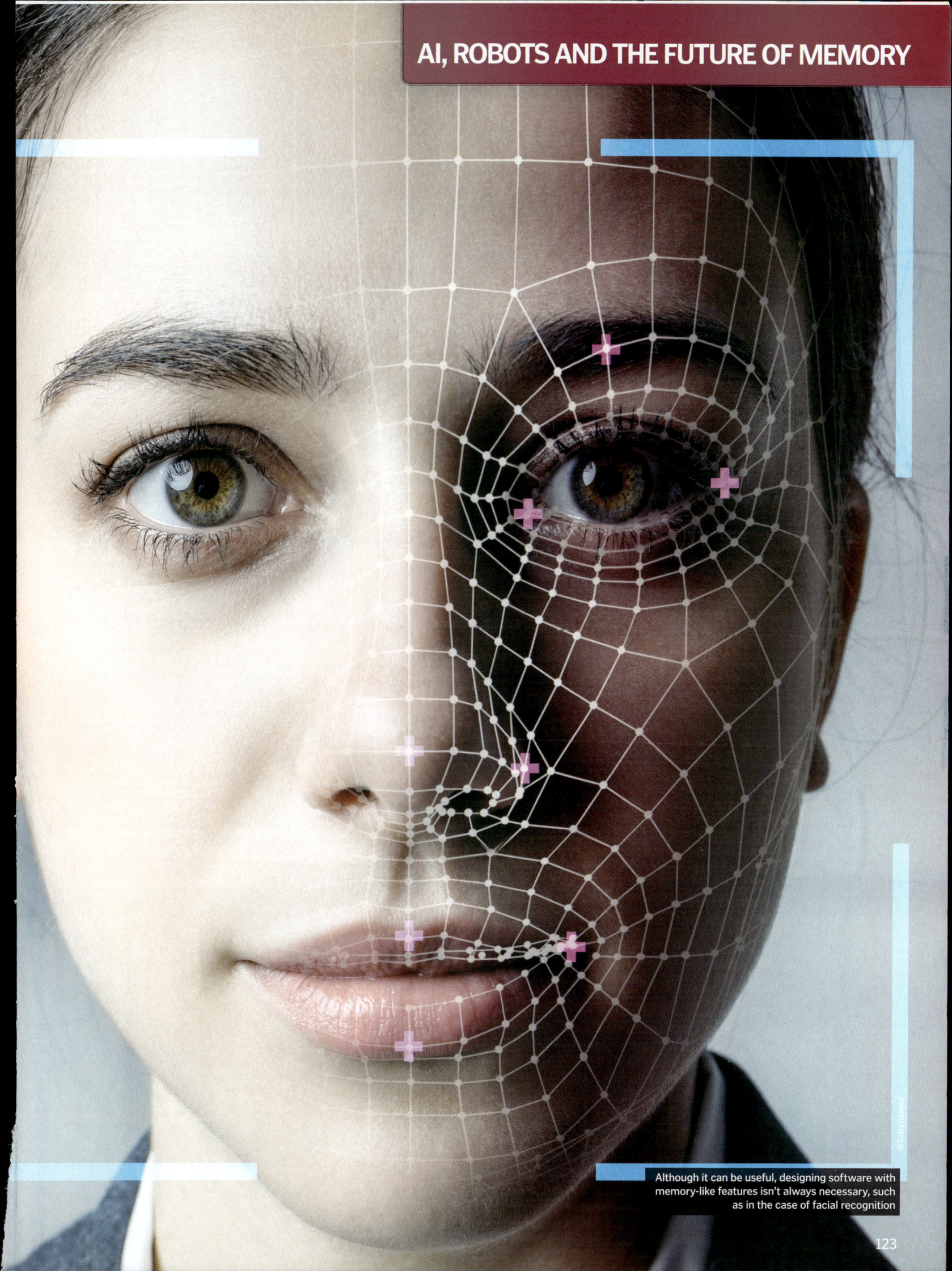

AI, ROBOTS AND THE FUTURE OF MEMORY

Although it can be useful, designing software with memory-like features isn't always necessary, such as in the case of facial recognition

FLORA, FAUNA AND THE FUTURE

understood organic brain, and they need a computer powerful and efficient enough to store the swathes of knowledge.

One of the most popular forms of 'learning' algorithm is the neural network. Roughly based on the human brain, a neural network is arranged in layers of nodes that play the part of neurons. Once they've received an input, the nodes will only 'fire' to connected nodes if the input signal is strong enough, emulating the basic principle of our brain chemistry. Basic neural networks learn through a process known as a backpropagation. The algorithm will first be 'trained' to make the correct classification (such as "this is a dog, and this is a cat"). If the algorithm gets the classification wrong in training, the error is noted and passed backward into the neural network, forcing it to adjust how much weight it awards certain features. In the case of the two animals, the algorithm may learn to pay less attention to the colour of the coat and focus more on the shape of the animal's eyes, for example. Algorithms based on this principle have proven capable of recognising patterns in previously unseen data, provided they have large data sets to learn from.

Simple neural networks can be compared to teaching a long-term memory in that through lots of repetition the algorithm can learn to recognise familiar stimuli. However, such networks are not encoded to remember the period of learning. In our animal example, the final algorithm may confidently predict the appearance of a dog by solely using the proportions of its nose, mouth and whiskers, but it won't remember that it once found the eyes a useful marker as well. This is not an issue when classifying images – in fact, not having to worry about memory is actually an asset to many algorithms.

The memory problem in computer science is a big one, primarily because having your algorithm remember what it did in every step quickly takes up a huge amount of storage space. With that being the case, designing an algorithm with no memory can work just fine. Imagine a humanoid robot with the assigned goal of ascending a snaking, uphill route to the top of a hill. In one action the robot is designed to do the following: it first randomly chooses a direction, either forward, backward, left or right, then extends a foot to test the surface. If the direction goes uphill, the robot completes the step. If not, it withdraws to its current position. Even without remembering where it came from or what direction it tried, the robot will ascend to the top of the hill, although to a human observer it may prove painfully boring to watch.

Much of the inspiration for programming memory has come from the human brain. Consolidating memories while we sleep, for example, helped inspire the 'compressive transformer' code

We can, however, get the robot there faster by granting it a tiny bit of short-term memory. If the robot stepped forward and went uphill in its previous action, now it'll preferentially try to go forward again. If last time going forward didn't lead uphill, then it'll avoid trying to go forward until it's moved uphill in another direction. In a race against the robot with no memory at all, merely having a memory that's forgotten two steps later would be hugely beneficial. Plus, if the goal is simply to ascend to the top of the hill, then having the robot retain knowledge of every previous step is resoundingly inefficient. However, if our goal is to create a general artificial intelligence with human-like memory, we'll need to do better.

One sophisticated way of getting around the problem of remembering every piece of information is designing an algorithm that can recognise and remember only the important parts. Humans innately acquire this ability; as children, when we learn to tie our shoelaces, we remember the motions of wrapping the fabric into a bow, but we don't remember how the lace felt. In 1997, a paper written by scientists Sepp Hochreiter and Jürgen Schmidhuber introduced equations pertaining to 'long short-term memory', which provided such a memory trick in computer form.

'Long short-term memory' algorithms work by allowing the algorithm to decide on what information

> "Neural networks can be compared to a long-term memory… Through repetition the algorithm can recognise stimuli"

Google's DeepMind, led by founder and CEO Demis Hassabis, is at the forefront of designing computer models with advanced working memories

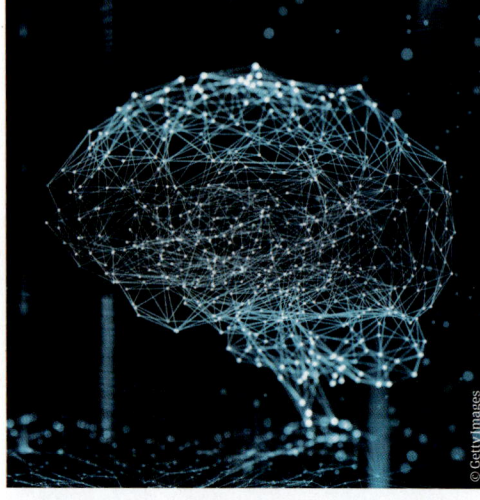

Neural networks are roughly based on the human brain, with layers of 'neurons' that are only activated when the signal feeding into them is strong enough

AI, ROBOTS AND THE FUTURE OF MEMORY

to remember, what parts to forget and when to forget them. A popular application of this technology has been in language modelling. When asking an algorithm to predict the next word in a sentence, having it remember just the previous five words is typically enough to generate a coherent speech. However, long short-term memory algorithms have been used to take this technology further, allowing the computer to remember an important piece of information and retain it until it ceases to be relevant and can be forgotten. As a simple example, an algorithm tasked with writing up a report for a football match would remember that the score was 0-0 until a team scored. At such a point, the score 1-0 would be remembered and 0-0 could be forgotten, as it would cease to have an impact on the text moving forward.

Researchers at Google's DeepMind, who are at the forefront of the artificial intelligence field, have subsequently improved upon the memory algorithm by once more turning to the human brain for inspiration. Their first improvement – which has also been pioneered by others in the field – is to encode 'attention' as well as working memory. In humans, our working memory is described as the ability to attain and access relevant information, while attention is to selectively process the information we encounter in the environment. DeepMind algorithms have been designed to use attention to be selective over the information retained by the model, helping to ensure that important information isn't lost and that excessive information isn't stored. This is especially useful when the scientists need to train their model with lots of different data. In these instances, even retaining all pieces of data for a brief period would require a massive memory.

In addition to this, another human-inspired forerunner technique used by DeepMind is the 'compressive transformer'. This code attempts to emulate the consolidation of memories that occurs as we sleep. With a compressive transformer in its code, a model will squish important bits of relevant information into a compressed form rather than discard them when they get older. The combination of these human-inspired mechanics is already bearing fruit: when shown an example of exceptional literary prose, a compression transformer-equipped algorithm has been able to recreate the quality faithfully. Similar models can even copy Shakespeare with high fidelity. Unlocking memory in computers is already paving the way to faithful reconstructions of our written word – an integral step in a world where we actively engage with computers every day. Such breakthrough technologies will also undoubtedly soon manifest in our virtual assistants, allowing us to enjoy natural-sounding conversations with our 'smart' machines.

What is simultaneously fascinating and frightening about these memory algorithms is that they can now become highly competent at fairly complex tasks without any understanding of what they're creating.

APPLICATIONS OF A MEMORY MACHINE

The main advantage of memory-based neural networks (also called recurrent neural networks) over their non-memory-storing cousins is that memory algorithms can handle sequential data. This means that they can recognise and predict patterns in a sequence (such as words) or through time. This second asset can be incredibly powerful for predicting the future based on 'memories' the algorithm has acquired from past events. Traffic flow, the weather and stresses acting on dams have all been accurately forecasted using similar models. However, the application of memory algorithms to predicting changes in the stock market may prove the most impactful over the foreseeable future.

The stock market – where shares in companies are constantly traded, inflating and deflating their value – is notoriously volatile and complex, making it hard to predict with any degree of accuracy by a human. However, memory-storing models have proven considerably more capable of accomplishing the task. For input data, which form the 'memories' of the model, scientists use previous stock market fluctuation trends as well as financial news. These input data have been used to inform algorithms based on long short-term memory, creating models that can accurately predict dips and rises in stock prices into the near future.

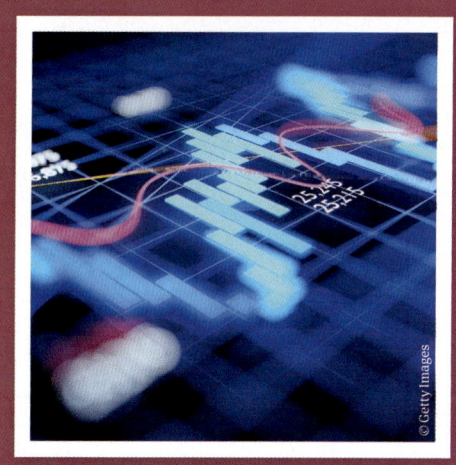

Memory-forming models have been frequently used to forecast changes in stock market prices

Deep neural networks have been used in software that's capable of completing complex tasks, such as defeating a world champion of the game Go

To an algorithm, the literary pieces it generates are simply formed by following logical, statistically driven inputs. The memories and the words themselves hold no meaning at all – the algorithm is simply following a recognised pattern. This realisation reveals both the potential and the limitations.

Synthetic memory is becoming possible, unlocking exceptional powers of computational mimicry. However, we remain far away from an algorithm that understands the meaning and power of its actions. Therefore, despite its incredible importance to the evolution and advancement of humankind, memory will form a comparatively much smaller step toward the creation of a general artificial intelligence.

TEST YOUR MEMORY

REMEMBER THE PAGE?

YOU'VE READ ALL ABOUT MEMORY, BUT HOW GOOD IS YOURS? CAN YOU REMEMBER WHERE YOU SAW THESE?

PAGE NUMBER:

....

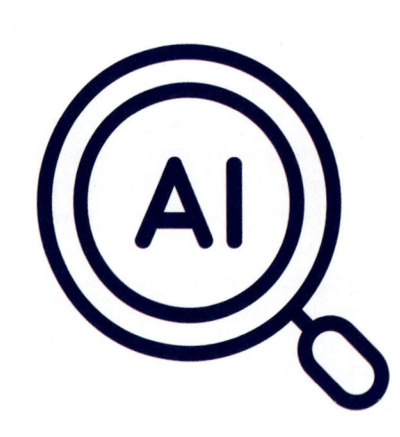

PAGE NUMBER:

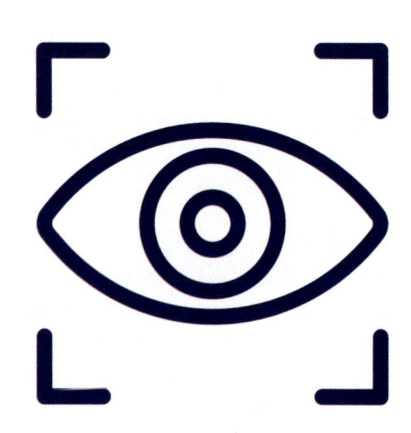

....

PAGE NUMBER:

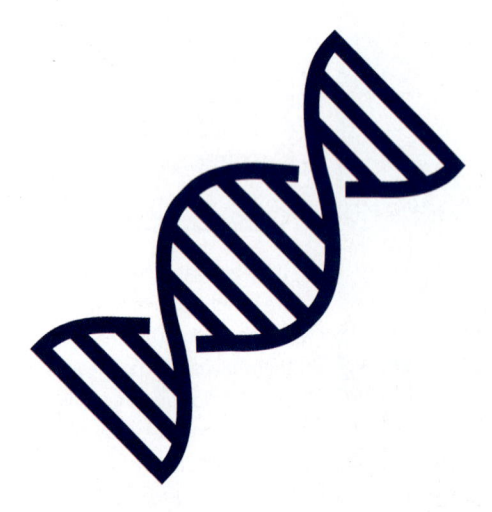

....

PAGE NUMBER:

....

PAGE NUMBER:

....

TEST YOUR MEMORY

MEMORY TEST

STUDY THE 20 WORDS BELOW, THEN CLOSE THIS BOOK AND SEE HOW MANY YOU CAN REMEMBER

MANDELA	TRAUMA	FORGET	MYSTERY	SAVANTS
BRAIN	RECALL	FACIAL	SLEEP	MALL
FALSE	PLANT	TEST	DOOR	TECHNOLOGY
DREAM	DOLPHIN	DEMENTIA	SUSIE	REMEMBER

QUICK-FIRE QUESTIONS

1. HOW MANY MEMORY FACTS WERE THERE IN THE OPENING ARTICLE?

....

2. WHAT IS THE SURNAME OF SUSIE, THE WOMAN WHO CAN'T REMEMBER?

....

3. IN WHICH ITALIAN CITY IS THERE A TRAIN STATION CLOCK MANY PEOPLE THINK HAS BEEN STUCK SINCE 1980?

....

4. CAN YOU NAME THE TWO PATIENTS WITH SEVERE MEMORY PROBLEMS WHO FEATURE IN THIS BOOK?

....

Answers: 1. 32. 2. McKinnon 3. Bologna 4. Henry and Eugene

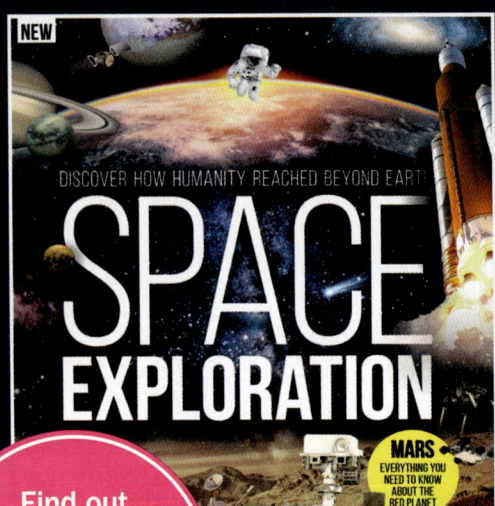

Find out everything you've ever wanted to know about outer space

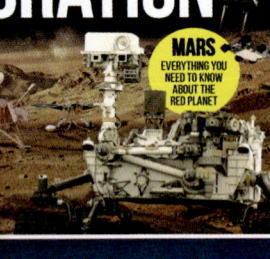

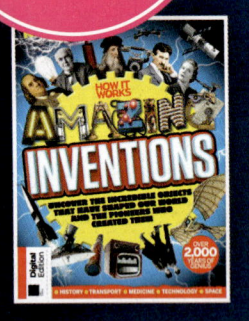

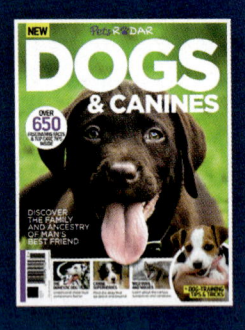

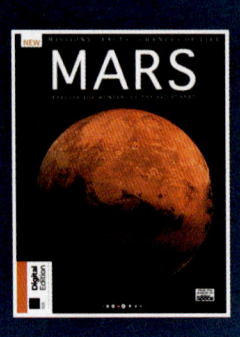

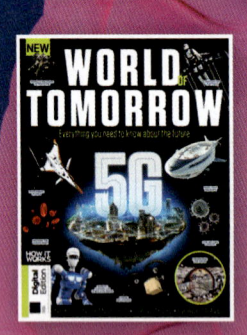

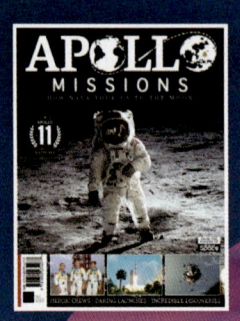

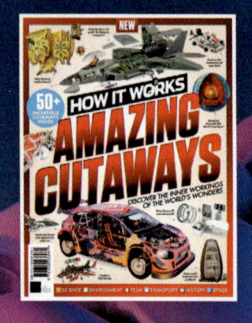

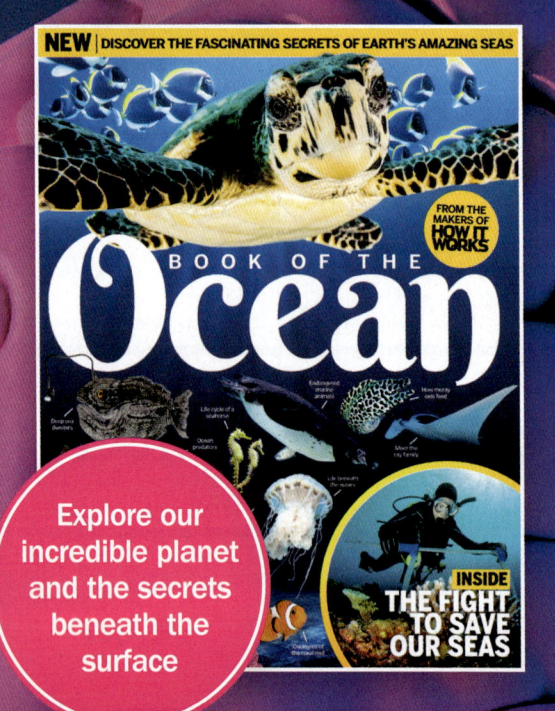

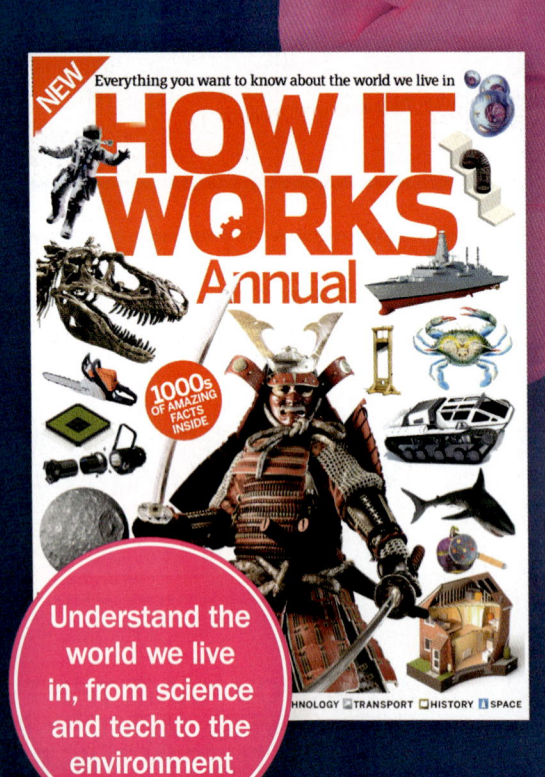

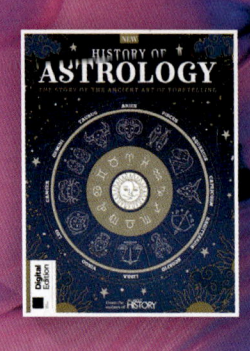

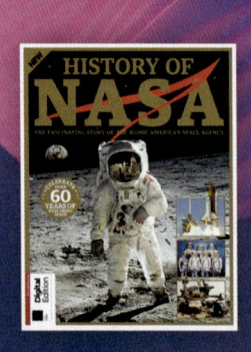

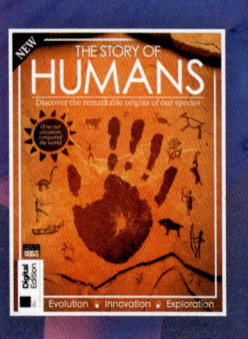

Explore our incredible planet and the secrets beneath the surface

Understand the world we live in, from science and tech to the environment

✓ Get great savings when you buy direct from us

✓ 1000s of great titles, many not available anywhere else

✓ World-wide delivery and super-safe ordering

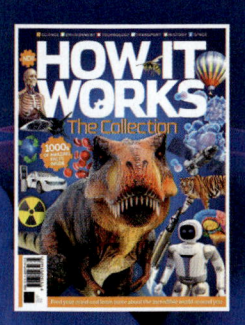

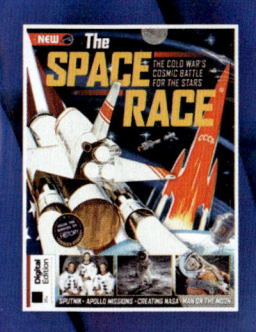

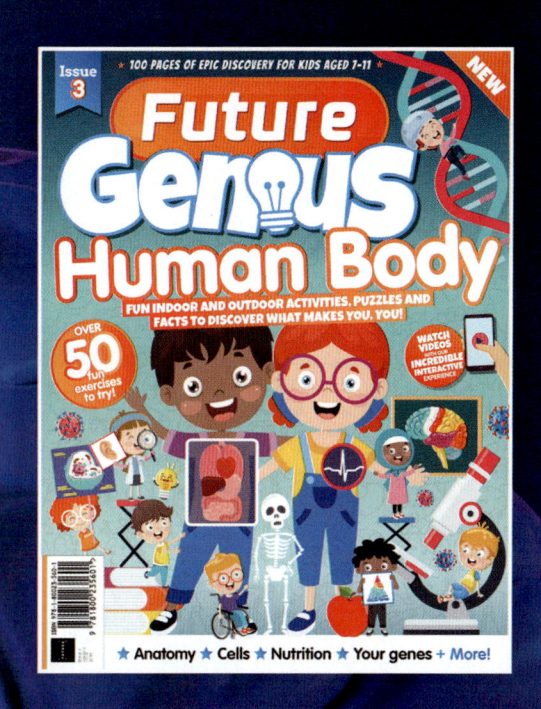

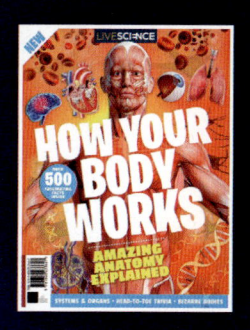

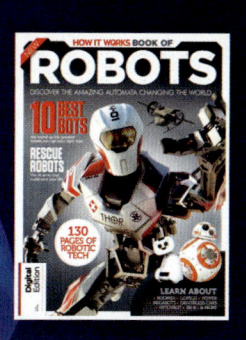

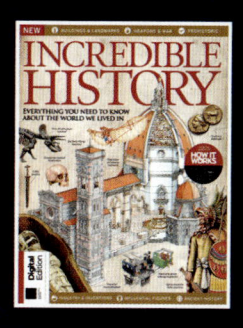

FEED YOUR MIND WITH OUR BOOKAZINES

Explore the secrets of the universe, from the days of the dinosaurs to the miracles of modern science!

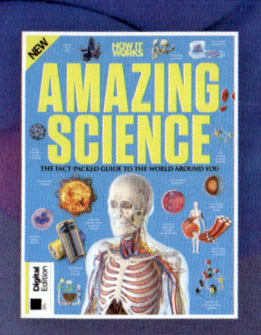

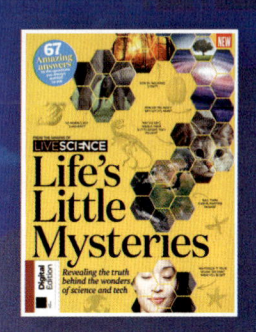

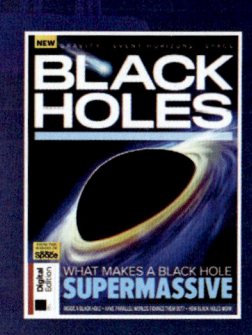

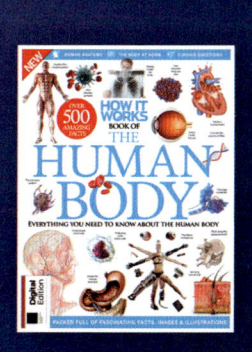

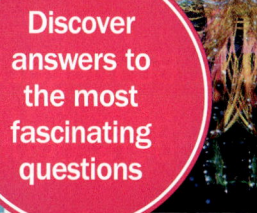

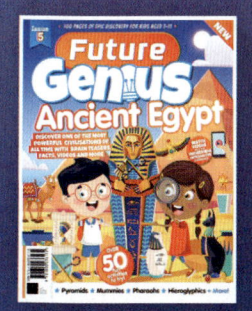

Discover answers to the most fascinating questions

Follow us on Instagram @futurebookazines

www.magazinesdirect.com
Magazines, back issues & bookazines.

FUTURE